AF564857

Evaluation and Impact Assessment

NIPA® GENX ELECTRONIC RESOURCES & SOLUTIONS P. LTD.
New Delhi-110 034

About the Authors

Dr. Arjun Prasad Verma completed his graduation degree from SVPUAT, Meerut, and Master's degree from Department of Agricultural Communication, GBPUAT, Pantnagar and PhD from Division of Dairy Extension, ICAR-NDRI, Karnal. He has been the recipient of ICAR-JRF (2013), UGC-JRF (2014) and ICAR-SRF (2015). He started his career as SMS (Agril. Ext.) at KVK, Bharari Jhansi and served 3.5 years. At present he is working as Assistant Professor at Department of Agricultural Extension Education, CoA, Banda University of Agriculture and Technology, Banda. His innovative research has resulted in more than 37 publications in journals of international and national repute. Besides, 39 popular articles, 3 Book, 04 chapters in books, 2 winter school, 1 Farmers fair, 2 National seminar, 7 News Letters, 27 Abstract, 3 Radio Talk, 7 e-reading manual, 5 AgMOOC courses and more than 50 newspapers coverage. He has awarded with 3 best paper presentation awards at national level. His area of specialization is Educational module & ICT application in agriculture and allied sciences. He has been serving as life time member of ISEE, New Delhi, SEE, Agra and RSEE, Rajasthan.

Dr. Pankaj Kumar Ojha He did his M.Sc. (Ag.) in Extension Education from Banaras Hindu University, Varanasi and Ph.D. in Extension Education from Dr. Rajendra Prasad Central Agricultural University, Pusa. He got University Gold medal in Ph.D. Besides these, he did Post graduate diploma in rural development from IGNOU, Post graduate diploma in Theatre communication from Banaras Hindu University, Varanasi and MBA (HR & IT) through distance learning mode from Vivekananda Subharti University, Meerut, U.P and PGDETM from University of Hyderabad. He has numbers of publications to his credit including edited books (Four), research papers, book chapters, review articles, popular articles, etc. He is Life member of various professional societies. He was also awarded the national award AIASA Gold Medal for his literary contribution. Presently he is working as an Assistant Professor, Department of Agricultural Extension, Banda University of Agriculture and Technology, Banda (U.P.).

Dr. Dheeraj Mishra is Assistant Professor of Agricultural Extension at Banda University of Agriculture and Technology, Banda. He received his M.Sc. (Ag.) Agricultural Extension degree from R.B.S. College, Bichpuri, Agra and Ph.D in Extension Education from Banaras Hindu University, Varanasi. He also earned PG Diploma in Mass Communication, PG Diploma in Educational Technology Management and Diploma in Theatre Communication. He has more than 25 research papers and 6 books in his account.

Dr. Brijesh Kumar Gupta, completed graduation degree, B.Sc (Agriculture) from Purvanchal University, Jaunpur, UP, M.Sc. (Agriculture) & PhD in Extension Education from the Department of Extension Education, Institute of Agricultural Sciences, Banaras Hindu University, Varanasi, Uttar Pradesh. He started his career as Cane Supervisor in State Sugarcane Development Department at Maharajganj District of Uttar Pradesh and served ten years. After that, he has joined Assistant Professor in Department of Agricultural Extension at College of Agriculture, Banda University of

Agriculture and Technology, Banda, Uttar Pradesh, India. Dr. Gupta has more than nine years' of experience in teaching, research and extension. He has also completed degree Master of Business Administration (MBA) from Sikkim Manipal University, Sikkim. He wrote 3 books, 18 research papers, 2 review papers, 12 popular articles and bulletins, 6 e-reading manual, 3 training manuals, 11 book chapters, 4 newsletters, 25 abstract etc. At present he is working as Assistant Professor in Department of Agricultural Extension at College of Agriculture, Banda University of Agriculture and Technology, Banda, Uttar Pradesh.

Dr. Bhanu Prakash Mishra completed Bachelors from DD Gorakhpur University, Gorakhpur, M.Sc. (Ag.) from ANDUAT, Kumarganj, Ayodhya and PhDfrom Department of Extension Education, Institute of Agricultural Sciences, Banaras Hindu University, Varanasi, U.P. He started his career as SMS (Agricultural Extension) at Krisi Vigyan Kendra, Khawzawl, Champhai district, Mizoram and serve there for 2 years. After that he has joined as an Assistant Professor (Extension Education and Rural Sociology) at College of Horticulture and Forestry, Central Agricultural University, Pasighat, Arunachal Pradesh, India. Dr. Mishra has 13 years' experience in teaching, research and extension education. He has written number of research papers, review papers, popular articles, bulletins, book chapters, etc. At present he is working as Professor and Head, Department of Agricultural Extension, College of Agriculture, Banda University of Agriculture and Technology, Banda, U.P.

Evaluation and Impact Assessment

Arjun Prasad Verma
Pankaj Kumar Ojha
Dheeraj Mishra
Brijesh Kumar Gupta
Bhanu Prakash Mishra

NIPA® GENX ELECTRONIC RESOURCES & SOLUTIONS P. LTD.
New Delhi-110 034

NIPA® GENX ELECTRONIC RESOURCES & SOLUTIONS P. LTD.

101,103, Vikas Surya Plaza, CU Block
L.S.C. Market, Pitam Pura, New Delhi-110 034
Ph : +91-11-43860225, Mob.: +91 9717133558, 9540816132
E-mail: newindiapublishingagency@gmail.com
Website: www.nipaersources.com

Print ISBN: 978-93-58879-90-2

ebook ISBN: 978-93-58872-35-4

Composed and Designed by NIPA®.

Preface

Evaluation has become an essential component of effective programme planning and implementation across development, education, agriculture, health, and allied sectors. In today's knowledge-driven society, the success of any programme is measured not only by its execution but also by its outcomes, impacts, and long-term sustainability. Recognizing this need, this book, Programme Evaluation, has been prepared to provide learners with both theoretical foundations and practical approaches to evaluation.

The book is organized into five blocks for systematic learning. The first block introduces the fundamental concepts and theories of evaluation. The second block explains the process of conducting evaluations in real-world contexts. The third block focuses on management techniques such as SWOT analysis, bar charts, and networks, which aid in programme planning and decision-making. The fourth block presents programme evaluation tools and models, including the Logic Framework Approach (LFA). The final block deals with impact assessment, offering insights into indicators, methodologies, and specialized areas such as Environmental Impact Assessment (EIA).

This structured approach ensures that the book serves as a comprehensive academic resource for students, especially those pursuing agricultural extension, rural development, and social sciences, in alignment with contemporary curricula and ICAR guidelines. At the same time, it is equally useful for practitioners, policymakers, and researchers who seek to apply evaluation techniques for evidence-based decision-making.

It is our sincere hope that this book will not only build conceptual clarity but also enhance analytical and critical thinking skills. By combining theoretical perspectives with practical tools, it aims to empower learners and professionals to design, implement, and evaluate programmes effectively. Constructive suggestions and feedback from readers are most welcome, as they will guide the refinement of future editions.

Authors

Contents

Contents

Block 1: Programme Evaluation

Unit 1

Introduction to Evaluation

Concept of Evaluation: Meaning and concept in different contexts

The term "evaluation" originates from the Latin word "Valupure," which signifies the value attributed to a particular entity, concept, or action. Evaluation is a systematic process that involves the periodic comparison of actual outcomes and impacts against those initially planned or anticipated. It is designed to assess the overall worth of an endeavor, thereby facilitating learning and improvements for future initiatives. This process encompasses the structured and objective assessment of an ongoing or completed project, program, or policy, including its design, execution, and results. The primary objective of evaluation is to determine the relevance, achievement of objectives, efficiency, effectiveness, impact, and sustainability of a given intervention. Through this approach, evaluation seeks to provide credible and useful information that supports informed decision-making for stakeholders, including beneficiaries and donors. By integrating lessons learned from past experiences, evaluation enhances the strategic planning and implementation of future projects, ensuring continuous improvement and better resource utilization. Evaluation is an essential tool in both public and private sectors, allowing organizations to measure their performance against predefined benchmarks. It serves as a crucial mechanism for accountability, ensuring that resources are utilized effectively to achieve intended goals. Moreover, evaluation fosters transparency by offering objective insights into the successes and shortcomings of initiatives, thereby reinforcing stakeholder trust and engagement. There are several types of evaluation, each serving a distinct purpose. **Formative evaluation** is conducted during the early stages of a project or program to refine and enhance its design and implementation. This type of evaluation focuses on identifying potential challenges, assessing feasibility, and ensuring that objectives align with stakeholders' needs. **Summative evaluation**, on the other hand, occurs after a project's completion and aims to assess its overall effectiveness and impact. It provides a comprehensive review of outcomes and offers recommendations for future initiatives. Another critical category is **process evaluation**, which examines the implementation phase to determine whether the project or program is being executed as planned. This type of evaluation assesses operational efficiency, adherence to guidelines, and stakeholder involvement. **Impact evaluation** delves into the long-term effects of an intervention, measuring its influence on target populations and broader social, economic, or environmental factors. Lastly, **sustainability evaluation** focuses on assessing whether the benefits of a project or policy will persist beyond its initial funding period. The effectiveness of an evaluation largely depends on the methodologies employed. Both qualitative and quantitative approaches are utilized to gather comprehensive data. Quantitative methods involve statistical analysis, surveys, and structured assessments to measure performance metrics objectively. Qualitative methods, such as interviews, focus groups, and case studies, provide deeper insights into stakeholder experiences, perceptions, and contextual factors influencing outcomes. Evaluation is an indispensable process that enables organizations and policymakers to assess the value and effectiveness of projects, programs, and policies. By systematically analyzing design, implementation, and results, evaluation facilitates informed decision-making, enhances accountability, and promotes sustainable development. Its role in fostering continuous improvement ensures that resources are efficiently allocated and objectives are effectively achieved, ultimately contributing to greater societal impact.

According to OECD (1991) *An evaluation is an assessment, as systematic and objective as possible, of an ongoing or completed project, programme or policy, its design, implementation and results. The aim is to determine the relevance and fulfilment of objectives, developmental efficiency, effectiveness, impact and sustainability. An evaluation should provide information that is credible and useful, enabling the incorporation of lessons learned into the decision-making process of both recipients and donors.*

According to C.V. Good *"The process of ascertaining or judging the value or amount of something by use of a standard of standard of appraisal includes judgement in terms of internal evidence and external criteria."*

According to Miller, Gronlund and Linn *"Evaluation is a systematic process of collecting, analysing and interpreting information to determine the extent to which pupil's are achievement instructional objectives."*

Why Evaluation is Done

Ensuring planned results are achieved

i. Emphasis on checking progress towards the achievement of an objective.

ii. Improving and support management

iii. Generating shared understanding

iv. Generating new knowledge and support learning

v. Building the capacity of those involved
vi. Motivating stakeholders
vii. Ensuring accountability
viii. Fostering public and political support
ix. Management decision-making
x. Organisational learning
xi. Soliciting support for programmes

When Evaluation is Done?

The evaluation of a program or project is considered done when required the following:

1. **All Evaluation Criteria Have Been Assessed**
   1. The program or project has been measured against its objectives, KPIs, or success metrics.
   2. Data collection (quantitative and qualitative) is complete.
2. **Findings and Analysis Are Completed**
   1. Data has been analyzed, trends identified, and insights gathered.
   2. Strengths, weaknesses, opportunities, and risks (SWOT) have been examined.
3. **Recommendations and Conclusions Are Finalized**
   1. Lessons learned and areas for improvement are documented.
   2. Clear recommendations for future actions are made.
4. **Final Report is Prepared and Shared**
   1. The evaluation results are compiled into a report or presentation.
   2. Findings are shared with stakeholders, funders, or decision-makers.
5. **Stakeholder Feedback is Considered**
   1. The evaluation findings are discussed with relevant stakeholders.
   2. Adjustments or clarifications are made based on feedback.
6. **Action Plan is Developed (If Needed)**
   1. If the evaluation results suggest improvements, an action plan is created.
   2. Next steps, responsibilities, and timelines are defined.

Programme planning

Programme planning is a decision making process which involve critical analysis of the existing situation, problems and evaluation of the various possibilities and or alternatives to solve these problems and the selection of the relevant ones which are most appropriate, giving necessary priorities based upon local needs and resources by the cooperative efforts of the people both official and non-official with a view to facilitate the individual and community growth and development. In program planning we are required to know where we are now, and where we ought to go, so that we may better judge what to do and how to do it? It prepares the basis for a course of future action. The word 'programme' has several distinct meanings in the dictionary. Programme planning means a proclamation, a prospectus, a list of events, a plan of procedure, a course of action prepared or announced beforehand, a logical sequence of operations to be performed in solving a problem. When used by an organization, it means a prospectus or a statement issued to promote understanding and interest in an enterprise. According to Kelsey and Hearne (1949), an "extension programme" is a statement of situation, objectives, problems and solutions'. According to the USDA (1956), an "extension programme" is arrived at co-operatively by the local people and the extension staff and includes a statement of:

- The situation in which the people are located;
- The problems that are a part of the local situation;
- The objectives and goals of the local people in relation to these problems; and
- The recommendations or solutions to reach these objectives on a long-time basis (may be several years) or on a short-time basis (may be one year or less).

Programme is a written statement containing a more pertinent factual data used in decision-making, the problems agreed upon with priority assignment and the possible solutions to the problems'.

Plan or Plan of work is an outline of activities so arranged as to enable efficient execution of the entire programme. It answers the questions of what, why, how, when, where and by whom the work is to be done.

Project is a single item of the annual plant containing the method of solution of a single selected problem.

Calendar of work is a plan of work arranged chronologically, according to the time when step of work is to be done. It is a time schedule of work.

Aim is a broad objective. It is a generalized statement of direction and may have several objectives. It is also said to be an end in view to give direction to the creative process.

Objective is a direction of movement. A well stated objective is always measurable. It is also said to be a goal of growth.

Goal is a distance in any given direction, proposed to be covered in a given time.

Principles of programme planning

After a critical analysis of the programme planning principles available in extension literature, Sandhu (1965) identified a set of principles that may be applicable in developing countries.

I. Programme

1. **Extension programme planning is based on analysis of the facts in a situation:** It is important to take into account the conditions that exist at a particular time. This implies that factors such as land, crops, economic trends, social structure, economic status of the people, their habits, traditions and culture, in fact, everything about the area in which the job is to be done and its people, may be considered while planning an extension programme for an area. These factors may be viewed in terms of established long-term objectives and rural policy. The outcome of previous plans should also be reviewed and results utilized. Brunner and Yang (1949) argue that there is no greater mistake than to assume that technical know-how alone will solve the problems of the farmers. They say that no programme or even technique can achieve the desired results when not in harmony with the culture of the people. 'Extension knows, if need be, the surer way is to effect cultural change by the slow but certain process of education'.
2. **Extension programme planning selects problems based on people's interests and needs:** Sound programme building selects problems based on people's needs. I t is necessary to select these problems which are most urgent and of widest concern. Choice of problems must be from among those highlighted by an analysis of the facts regarding what are felt as unfelt needs. To be effective, extension work must begin with the interests of the families. It must meet interest and use them as a spring-board for developing further interests. It is common knowledge that people join together because of mutual interests and needs. Brunner (1945) said that an extension programme must meet the felt needs of the people. Leagans (1961) has recommended that the extension workers adopt the subject matter and teaching procedure to the educational level of the people, to their needs and interests, and to their resources.

3. **Extension programme planning determines definite objectives and solutions which offer satisfaction**: In order to hold interest, we must set working objectives and offer solutions which are within reach and which will give satisfaction on achievement. This is related to motivation for action. People must see how they or their communities are going to benefit from the proposed solutions. Very often the simplicity or dramatic effect of the practice recommended is the most potent factor in its wide adoption. Further, if there is to be progress and not more evolution in the development of man, the objectives must be periodically revised in view of the progress made. In other words, as changes occur, objectives need to be re determined to allow for even further progress to be realized.
4. **Extension programme planning has permanence with flexibility:** Any good programme must be forward looking and permanent. Permanence means anticipating years of related and well organized effort. Along with this lower process, which both follow and makes a long-term trend, experience has shown that particular items will need to be changed to meet unforeseen contingencies or emergencies. Without flexibility, the programme may not, in fact, meet the needs of the people. A programme should be prepared well in advance of its execution but not too far ahead of time. Ordinary events may subject it to change in part though not in total. It is obvious that an extension programme must be kept flexible to meet the changing needs and interests of the people.
5. **Extension programme planning has balance with emphasis**: A good programme should cover the majority of people's important interests. It must be comprehensive enough to embrace all groups, creeds and races at all levels and community, block, state, national and international problems. It is futile to deal with only one phase of life in a community as an end in itself. At the same time, a few of the most important or timely problems should be chosen for emphasis. To avoid scattered effort, something must stand out. Decisions must be made as to which of the needs are most urgent. The next consideration in choosing items for emphasis is to promote efficiency by permitting a good distribution of time and effort throughout the year. Too many things carried out simultaneously will divide either the worker's or the people's attention.

II. Planning process

6. **Extension programme planning has a definite plan of work**: No matter how well a programme is thought through, it is of no use unless carried out. This implies good organization and careful planning for action. A plan of work is an outline of procedure so arranged as to enable

efficient execution of the entire programme. It is the answer to what, where, when and how the job will be done. In carrying out programme plans, different leaders and groups may work on various phases, i.e., the women in the community may work on one segment, the men on a second segment and youth-club members on a third. Organization should be used as a tool to accomplish these purposes, never as an end in itself.

7. **Extension programme planning is an educational process**: The people who do the planning may participate in local surveys and neighbourhood observations. This provides an opportunity for them to learn more about their own community and area and increases their interest. The extension worker has the responsibilities of providing local leaders with the knowledge, skills and attitudes they must have if they are to help in educationally serving the people. Essentially, learning takes place through the experiences the learner has and the responses he makes to the stimuli of his environment. The experience gained in finding facts, analyzing situations, recognizing problems, stating objectives and thinking of possible solutions and alternatives should make for a better and more effective learning environment. The extension personnel should remember this fact and provide opportunities for the effective participation of local people in programme planning.

8. **Extension programme planning is a continuous process**: Since programme planning is viewed as an educational process and since education is seen as a continuous process, therefore it logically holds that extension programme planning is a continuous process. There is no question of exhausting new knowledge, either in the subject matter with which we deal or in the methods of teaching. With the constant flux of agricultural technology, extension education is faced with an increasingly more difficult job as it tries to serve the needs and interests of the people. Sutton (1961) said that extension in a changing society must adjust and plan for the future to serve the needs of people. He set forth five steps within might be useful in making necessary adjustments:

   1. Keep choice to the people
   2. Be flexible and ready to grasp with firmness new problems as they arise.
   3. Work with people in seeking practical solutions to their problems.
   4. Keep abreast of technological and social change.
   5. Close the gap between research discovery and practical application.

9. **Extension programme planning is a coordinating process**: Extension programme planning finds the most important problems and seeks agreement on definite objectives. It coordinates the efforts of all interested leaders, groups and agencies and considers the use of resources. It obtains the interest and co-operation of many people by showing them why things need to be done. This is important in working with people.
10. **Extension programme planning involves local people and their institutions:** Involvement of local people and their institutions is very essential for the success of any programme for their development. People become interested and give better support to the programme when they are involved in the planning process. So, extension programmes should be planned with the people and not for them.
11. **Extension programme planning provides for evaluation of results** Since extension programme planning involves decision-making procedures, so evaluation is important in order to make intelligent decisions aimed at achieving the stated objectives.

Step involved in Programme Planning

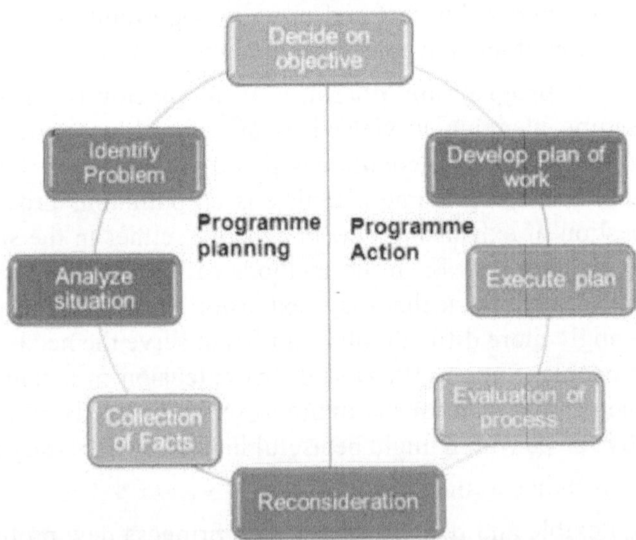

Source: http://ecoursesonline.iasri.res.in/mod/resource/view.php?id=4384

What Is Decision making

Decision making refers to making choices among alternative courses of action—which may also include inaction.

Types of Decisions

i. **Programmed decisions;** these are decisions that occur frequently enough that we develop an automated response to them.

ii. Decisions that are unique and important require conscious thinking, information gathering, and careful consideration of alternatives. These are called **nonprogrammed decisions.**

The levels of decision-making in a program or project evaluation typically fall into three main categories

1. **Strategic Level (Top-Level Decisions)**
   - **Who decides?** Top management team, CEOs, Board of Directors, Senior executives, board members, policymakers, funders.
   - **Focus:** Long-term impact, policy changes, funding allocation, program continuation or termination.

   Examples
   - Should the program continue, expand, or be terminated?
   - Should more resources be allocated?
   - What long-term strategic adjustments are needed?
2. **Tactical Level (Mid-Level Decisions)**
   - **Who decides?** Project managers, department heads, supervisors.
   - **Focus:** Operational efficiency, resource optimization, mid-term adjustments.

   Examples
   - Should we change the project timeline or budget?
   - Should we adjust the implementation strategy based on evaluation findings?
   - What improvements should be made to processes?
3. **Operational Level (Day-to-Day Decisions)**
   - **Who decides?** Team leads, field staff, project coordinators.
   - **Focus:** Day-to-day execution, immediate problem-solving, minor adjustments.

   Examples
   - How should tasks be adjusted based on evaluation feedback?
   - What training or support do team members need?
   - How should we modify data collection or reporting methods?

Steps in Decision-Making

A well-structured decision-making process ensures that choices are made logically and effectively, especially in program or project evaluation. Below is a detailed 10-step framework for making informed decisions:

1. **Identify the Problem or Decision to Be Made**
   - Clearly define the issue or decision that needs to be addressed.
   - Ensure alignment with organizational goals, program objectives, or stakeholder needs.
   - Example: Should we expand, modify, or discontinue the program?
2. **Establish Decision Criteria**
   - Determine what factors will influence the decision (e.g., cost, feasibility, impact, sustainability).
   - Rank criteria based on priority (e.g., effectiveness, efficiency, stakeholder acceptance).
   - Example: If cost is the primary concern, financial feasibility might be a key criterion.
3. **Collect and Analyze Relevant Information**
   - Gather qualitative and quantitative data from the program evaluation.
   - Use stakeholder feedback, performance reports, financial data, and risk assessments.
   - Consider best practices and benchmark against similar projects.
4. **Identify Possible Alternatives**
   - Brainstorm different courses of action.
   - Include innovative and practical solutions.
   - **Example:**
     1. Continue the project as is.
     2. Modify certain aspects to improve performance.
     3. Scale up or expand.
     4. Discontinue and redirect resources.
5. **Evaluate the Alternatives against the Criteria**
   - Compare each alternative based on the decision criteria.
   - Use decision matrices, cost-benefit analysis, SWOT analysis, or risk assessments.
   - Example: A decision matrix can score each alternative based on cost, impact, and feasibility.

6. **Weigh the Risks and Uncertainties**
    - Identify potential risks and obstacles.
    - Consider the likelihood and impact of risks for each alternative.
    - Develop risk mitigation strategies where necessary.
7. **Choose the Best Alternative**
    - Select the most viable and effective solution based on the evaluation.
    - Ensure stakeholder consensus where necessary.
    - Validate the decision with expert consultation if required.
8. **Develop an Action Plan for Implementation**
    - Outline steps for execution, assign responsibilities, and allocate resources.
    - Set timelines, milestones, and success indicators.
    - Communicate the decision and next steps clearly to all stakeholders.
9. **Implement the Decision**
    - Execute the chosen alternative according to the action plan.
    - Monitor progress through feedback mechanisms and performance tracking.
10. **Monitor, Evaluate, and Adjust if Necessary**
    - Continuously track the effectiveness of the decision.
    - Conduct follow-up evaluations and stakeholder feedback sessions.
    - Make adjustments or refinements if new information arises.

Accountability

Accountability is the acceptance of responsibility for one's own actions. It implies a willingness to be transparent, allowing others to observe and evaluate one's performance.

Types of Accountability

- Corporate Accountability
- Political Accountability
- Government Accountability
- Media Accountability
- Accountability in the Workplace

Benefits of Accountability

- Accountability promotes operational excellence
- Accountability safeguards institute/organizational resources
- Accountability builds external investor trust

Impact Assessment

Impact Assessment (IA) is a systematic process used to evaluate the potential or actual effects of a project, policy, or program on society, the economy, and the environment. It helps decision-makers understand the implications of their actions and ensures that interventions achieve their intended outcomes while minimizing negative consequences.

What is the purpose of impact assessment?

According to the International Association for Impact Assessment (IAIA), we can break this down into four distinct goals:

- Understand the possible consequences of a proposed action, change, or intervention and plan ahead to respond to any positive and negative results.
- Encourage accountability to stakeholders, including shareholders, employees, donors, partners, customers, volunteers, and beneficiaries.
- Identify necessary procedures and methods for future policy, planning, and project cycles.
- Make environmentally, socially, and economically sustainable decisions for organizational growth and development.

Policy advocacy

Policy advocacy is a active, covert, or inadvertent support of a particular policy or class of policies. It can include a variety of activities including, lobbying, litigation, public education, and forming relationships with parties of interest. Advocating for policy can take place from a local level to a state or central government. Policy advocacy is the process of negotiating and mediating a dialogue through which influential networks, opinion leaders, and ultimately, decision makers take ownership of your ideas, evidence, and proposals, and subsequently act upon them. At their core are a number of ideas that continually come up, characterizing policy advocacy as follows:

- **Astrategy to affect policy change or action**—an advocacy effort or campaign is a structured and sequenced plan of action with the purpose to start, direct, or prevent a specific policy change.

- **A primary audience of decision makers**—the ultimate target of any advocacy effort is to influence those who hold decision-making power. In some cases, advocates can speak directly to these people in their advocacy efforts; in other cases, they need to put pressure on these people by addressing secondary audiences (for example, their advisors, the media, the public).
- A **deliberate process of persuasive communication**—in all activities and communication tools, advocates are trying to get the target audiences to understand, be convinced, and take ownership of the ideas presented. Ultimately, they should feel the urgency to take action based on the arguments presented.
- **A process that normally requires the building of momentum and support behind the proposed policy idea or recommendation.** Trying to make a change in public policy is usually a relatively slow process as changing attitudes and positions requires ongoing engagement, discussion, argument, and negotiation.
- **Conducted by groups of organized citizens**—normally advocacy efforts are carried out by organizations, associations, or coalitions represent the interests or positions of certain populations, but an individual may, of course, spearhead the effort.

Evaluation principles; the context of program evaluation in agricultural extension

Evaluation is a systematic process of determining to what extent instructional objectives has been achieved. Therefore evaluation process must be carried out with effective techniques.

The following principles will help to make the evaluation process an effective one

i. It must be clearly stated what is to be evaluated
ii. A variety of evaluation techniques should be used for a comprehensive evaluation
iii. An evaluator should know the limitations of different evaluation techniques
iv. The technique of evaluation must be appropriate for the characteristics or performance to be measured
v. Evaluation is a means to an end but not an end in itself

Functions of Evaluation

The function of evaluation process can be summarized as following:

i. Evaluation helps in preparing instructional/programme/project objectives
ii. Evaluation process helps in assessing the stakeholders needs
iii. Evaluation help in providing feedback to the stakeholders
iv. Evaluation helps in preparing programmed materials
v. Evaluation helps in reporting progress of the project/programme
vi. Evaluation data are very much useful in guidance and planning of future programme/project.
vii. Evaluation helps in effective administration
viii. Evaluation data are helpful in research

Role of Evaluator

Evaluator can play their role in several ways that is given below;

i. Role as educator
ii. Role as facilitator
iii. Role as consultant
iv. Role as interpreter
v. Role as mediator
vi. Role as change agent

Competency of evaluator

Competencies can be defined as "clusters of related knowledge, skills, abilities and other requirements necessary for successful job performance." The competency of evaluator has been reflected in various domains which are given below;

A. **Professional Practice/Foundation:** focuses on what makes evaluators distinct as practicing professionals. The competent evaluator;

   i. Acts ethically through evaluation practice that demonstrates integrity and respects people from different cultural backgrounds and indigenous groups.
   ii. Applies the foundational documents adopted by the American Evaluation Association that ground evaluation practice.
   iii. Selects evaluation approaches and theories appropriately.
   iv. Uses systematic evidence to make evaluative judgments.

v. Reflects on evaluation formally or informally to improve practice.
vi. Identifies personal areas of professional competence and needs for growth.
vii. Pursues ongoing professional development to deepen reflective practice, stay current, and build connections.
viii. Identifies how evaluation practice can promote social justice and the public good.
ix. Advocates for the field of evaluation and its value.

B. **Methodological Competency:** focuses on technical aspects of evidence-based, systematic inquiry for valued purposes. Methodology includes quantitative, qualitative, and mixed designs for learning, understanding, decision making, and judging. The competent evaluator;
i. Identifies evaluation purposes and needs.
ii. Determines evaluation questions.
iii. Designs credible and feasible evaluations that address identified purposes and questions.
iv. Determines and justifies appropriate methods to answer evaluation questions, e.g., quantitative, qualitative, and mixed methods.
v. Identifies assumptions that underlie methodologies and program logic.
vi. Conducts reviews of the literature when appropriate.
vii. Identifies relevant sources of evidence and sampling procedures.
viii. Involves stakeholders in designing, implementing, interpreting, and reporting evaluations as appropriate. 2.9 Uses program logic and program theory as appropriate.
ix. Collects data using credible, feasible, and culturally appropriate procedures.
x. Analyzes data using credible, feasible, and culturally appropriate procedures.
xi. Identifies strengths and limitations of the evaluation design and methods.
xii. Interprets findings/results in context.
xiii. Uses evidence and interpretations to draw conclusions, making judgments and recommendations when appropriate.

C. **Context Competency:** focuses on understanding the unique circumstances, multiple perspectives, & changing settings of evaluations and their users/stakeholders. The competent evaluator

i. Responds respectfully to the uniqueness of the evaluation context.
ii. Engages a diverse range of users/stakeholders throughout the evaluation process.
iii. Describes the program, including its basic purpose, components, and its functioning in broader contexts.
iv. Attends to systems issues within the context.
v. Communicates evaluation processes and results in timely, appropriate, and effective ways.
vi. Facilitates shared understanding of the program and its evaluation with stakeholders.
vii. Clarifies diverse perspectives, stakeholder interests, and cultural assumptions.
viii. Promotes evaluation use and influence in context.

D. **Planning & Management Competency:** focuses on determining and monitoring work plans, timelines, resources, and other components needed to complete and deliver an evaluation study. The competent evaluator;

i. Negotiates and manages a feasible evaluation plan, budget, resources, and timeline.
ii. Addresses aspects of culture in planning and managing evaluations.
iii. Manages and safeguards evaluation data.
iv. Plans for evaluation use and influence.
v. Coordinates and supervises evaluation processes and products.
vi. Documents evaluation processes and products.
vii. Teams with others when appropriate.
viii. Monitors evaluation progress and quality and makes adjustments when appropriate.
ix. Works with stakeholders to build evaluation capacity when appropriate.
x. Uses technology appropriately to support and manage the evaluation.

E. **Interpersonal Competency:** focuses on human relations and social interactions that ground evaluator effectiveness for professional practice throughout the evaluation. Interpersonal skills include cultural competence, communication, facilitation, and conflict resolution. The competent evaluator;

i. Fosters positive relationships for professional practice and evaluation use.
ii. Listens to understand and engage different perspectives.
iii. Facilitates shared decision making for evaluation.
iv. Builds trust throughout the evaluation.
v. Attends to the ways power and privilege affect evaluation practice.
vi. Communicates in meaningful ways that enhance the effectiveness of the evaluation.
vii. Facilitates constructive and culturally responsive interaction throughout the evaluation.
viii. Manages conflicts constructively

Credibility of evaluator

Evaluator's credibility is the extent to which the evaluator of a program is viewed by the stakeholders as trustworthy, technically competent, and knowledgeable enough to execute an evaluation in a fair and responsible manner.

Unit 2

Evaluation Theories

Evaluation theories refer to the conceptual frameworks, models, and principles that guide the systematic assessment and analysis of programs, policies, interventions, and other social phenomena. They provide a set of organizing principles and methodologies for evaluating the effectiveness, efficiency, relevance, and sustainability of various interventions and initiatives in different domains, including education, healthcare, social services, environmental protection, and public policy. Evaluation theories draw from various disciplines, such as psychology, sociology, economics, statistics, and management, and they may emphasize different aspects of the evaluation process, such as the role of stakeholders, the criteria for success, the methods for data collection and analysis, and the use of evaluation results.

i. Programme theory

Program theory defines as the mechanisms that mediate between the delivery (and receipt) of the program and the emergence of the outcomes of interest. Programme theory, variously referred to as programme theory, programme logic, theory-based evaluation or theory of change, theory-driven evaluation, theory-of-action, intervention logic, impact pathway analysis, and programme theory-driven evaluation science refers to a variety of ways of developing a causal modal linking programme inputs and activities to a chain of intended or observed outcomes, and then using this model to guide the evaluation. The term 'logic model' is used to refer to the summarized theory of how the intervention works (usually in diagrammatic form) and 'programme theory evaluation' is used for the process of developing a logic model and using this in some way in an evaluation. Most approaches to building logic models have focused on simple, linear models, but some have explored how non-linear models might be used to better represent programmes and guide their evaluation. In particular, a number of evaluators have incorporated concepts of complexity in their discussion and use of logic models.

Simple Logic Models

Many logic models used in programme theory (and guides to developing programme theory) show a single, linear causal path, often involving some variation on five categories (inputs, processes, outputs, outcomes and impact). Figure 1 shows a particularly influential version of this in a guide to developing and using logic models published by the W. K. Kellogg Foundation. For some interventions, simple logic models such as these are quite appropriate. There is, however, currently considerable debate about how appropriate this is for many human service interventions such as education, drug prevention, family support services and international development.

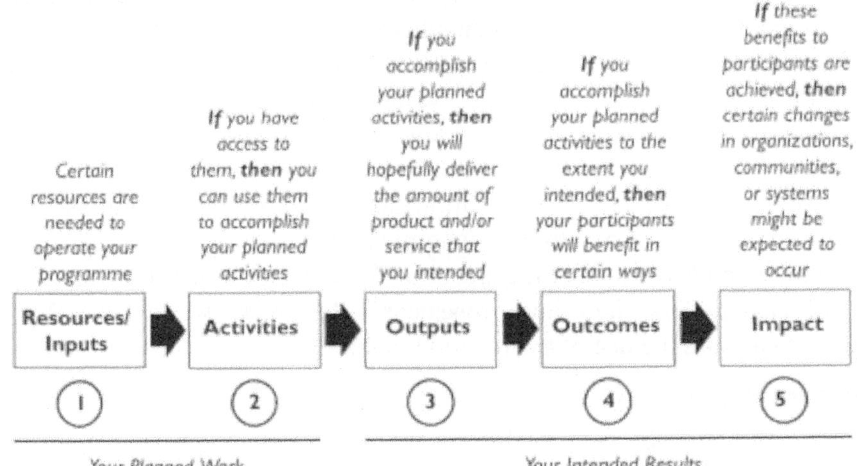

A Simple Logic Model (W. K. Kellogg Foundation, 2004)

Complicated Logic Models

Aspects of Complication

Three aspects of complication have been addressed in published examples of evaluations: interventions implemented through multiple agencies; interventions with multiple simultaneous causal strands; and interventions with alternative causal strands. Each of these will be discussed.

Multi-Site, Multi-Governance

One way in which interventions can be complicated, without necessarily being complex, is when they are implemented in different sites and/or under different governances. For example, the World Bank's (2004) account of efforts since 1974 to control the parasitic worm that causes onchocerciasis, or river blindness, in Africa refers to the partnership of eleven national governments, four international sponsors, and a host of private-sector companies and non-government organizations. Clearly, this presents challenges in terms of reaching agreement about evaluation planning, methods for data collection and analysis, and reporting to different audiences. However, there was a good understanding of the causal path for the intervention. Scientists knew which parasitic worm caused the problem and how its lifecycle could be interrupted. Although there was some local adaptation of implementation, there was an overall plan that was understood and agreed.

Aspect of complication	**Simple intervention**	**Complicated intervention**
1. Governance and location	Single organization	Multiple agencies, often interdisciplinary and cross-jurisdictional
2. Simultaneous causal strands	Single causal strand	Multiple simultaneous causal strands
3. Alternative causal strands	Universal mechanism	Different causal mechanisms operating in different contexts

Three Aspects of Complication

Simultaneous Causal Strands

A second aspect of complication is the existence of two or more simultaneous causal strands that are all required in order for the intervention to succeed. It can be important for a logic model to show both of these, and for the evaluation to gather data about both of them, so that important parts of the intervention are not left out. Simultaneous causal strands were an important aspect addressed in the monitoring and evaluation of a maternal and child health service. The programme sought to support parents to develop confidence in parenting at the same time as trying to encourage them to adopt healthier nutritional practices. Programme staff believed it was important to show in the logic model the need for balance between these two causal paths, to explain why they could not simply focus on strong messages about nutrition. It was therefore seen as necessary to show two different causal strands that were sometimes in tension–a strand where the expert staff provided skills and knowledge to the parents; and a strand where staff supported parents to have confidence in their parenting abilities. This was an important point when using the logic model as a communication device to explain the intervention to new staff or to other agencies, and when deciding which aspects should be included in monitoring or evaluation. To make it clear that these causal strands are not optional alternatives but each essential, it might be better to represent them using arrows that show clearly that both are required.

An intervention with multiple simultaneous causal strands cannot afford to only focus on achieving one of these. An evaluation needs to both document and support this. The previous example of controlling river blindness has something of this element as well. The success of the intervention depended on achieving success in each of the different locations (otherwise worms would simply breed in some areas and reinfest the others). It may have therefore been important to show simultaneous causal strands for each of the sites, all of which were necessary to achieve the final outcome.

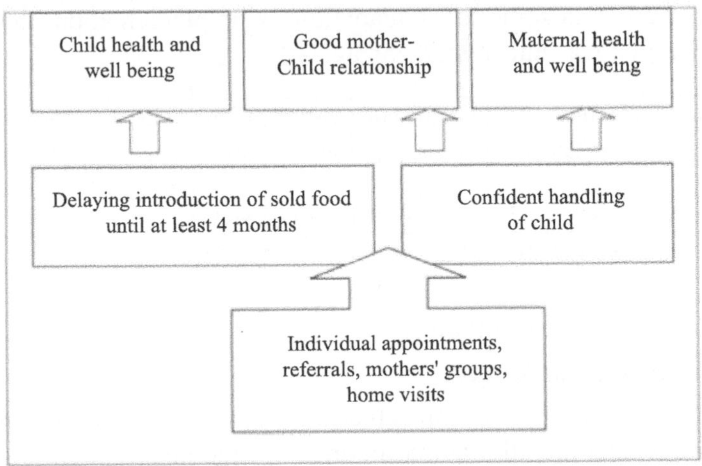

Logic Model Showing Simultaneous Causal Strands

Alternative Causal Strands

A third aspect of interventions that can be considered complicated involves alternative causal strands. In most logic models, these appear as parallel lines of boxes and arrows and are visually indistinguishable from simultaneous causal strands. The difference is that a programme can work either through one or other of the causal paths. In many cases, these alternative causal strands are effective in particular contexts–the 'what works for whom in what ways' notion of realist evaluation. So closed circuit television may work to reduce automobile theft in different ways in different contexts – through passive surveillance where it increases usage of car parks with spare capacity, or through capturing and removing thieves in contexts where it is part of an enforcement programme and where there is not a large pool of potential thieves to fulfill the void once some are removed.

Complex Interventions and Logic Models

Two Aspects of Complication

Aspect of complexity	Simple intervention	Complex intervention
1. Recursive causality and disproportionate effect	Linear, constant dose-response relationship	Recursive, with feedback loops, including reinforcing loops; disproportionate effects at critical limits
2. Emergent outcomes	Pre-identified outcomes	Emergent outcomes

i. Social science theory

ii. Evaluation theory

iii. Utilization-Focused Evaluation (U-FE)

Michael Quinn Patton, PhD, developed Utilization-Focused Evaluation (UFE) on the premise that "evaluations should be judged by their utility and actual use" (Patton, 2013). This theoretical model should be applied when the end goal is instrumental use (i.e. discrete decision-making). UFE focuses on intended use by *primary intended users*. To engage primary intended users, the evaluator must identify stakeholders who have the most direct, identifiable stake in the evaluation and its results, in other words, the "personal factor" (Patton 2013). The evaluator involves intended users at every stage of the process. The ultimate purpose of UFE is programmatic improvement driven by a psychology of use. Intended users are more likely to use the evaluation if they feel ownership of the process and its results. Use does not happen naturally; therefore, the evaluator must reinforce utility by engaging intended users at each stage of the evaluation. Patton prescribes a 17 step process for facilitating UFE from beginning to end.

The checklist is based on Essentials of Utilization-Focused Evaluation (Patton, 2012, Sage Publications).

Step 1: Assess and build program and organizational readiness for utilization-focused evaluation.

Step 2: Assess and enhance evaluator readiness and competence to undertake a utilization focused evaluation.

Step 3: Identify, organize, and engage primary intended users.

Step 4: Conduct situation analysis with primary intended users

Step 5: Identify primary intended uses by establishing the evaluation's priority purposes.

Step 6: Consider and build in process uses if appropriate.

Step 7: Focus priority evaluation questions.

Step 8: Check that fundamental areas for evaluation inquiry are being adequately addressed.

Step 9: Determine what intervention model or theory of change is being evaluated.

Step 10: Negotiate appropriate methods to generate credible findings and support intended use by intended users.

Step 11: Make sure intended users understand potential controversies about methods and their implications.

Step 12: Simulate use of findings.

Step 13: Gather data with ongoing attention to use.

Step 14: Organize and present the data for use by primary intended users.

Step 15: Prepare an evaluation report to facilitate use and disseminate significant findings to expand influence.

Step 16: Follow up with primary intended users to facilitate and enhance use.

Step 17: Meta-evaluation of use: Be accountable, learn, and improve

v. Values Engaged Evaluation Theory

Jennifer Greene, PhD, developed Values Engaged Evaluation (VEE) as a democratic approach that is highly responsive to context and emphasizes stakeholder values. VEE seeks to provide contextualized understandings of social programs that have particular promise for underserved and underrepresented populations. It is considered a "democratic" approach because it encourages the evaluator to include all relevant stakeholder values. Greene offers three justifications for including stakeholder values: (1) pragmatic (i.e. increases chance of use), (2) emancipatory (i.e. empowers stakeholders), and (3) deliberative (i.e. considers all interests). With this approach, evaluation design and methodology evolves as the evaluator understands the context, needs, and values underlying the program. VEE is concerned with answering broad and in-depth questions, and is more suited for formative rather than summative evaluations.

vi. Empowerment Evaluation Theory

David Fetterman, PhD, developed Empowerment Evaluation as an approach to foster program improvement through empowerment and self-determination. Self-determination theory describes an individual's agency to chart his or own course in life and the ability to identify and express needs. Fetterman believes the evaluator's role is to empower stakeholders to take ownership of the evaluation process as a vehicle for self-determination. The evaluator engages a diverse range of program stakeholders and acts as a "critical friend" or "coach" while guiding them through the evaluation process. Empowerment evaluation seeks to increase the probability of program success by providing stakeholders with the tools and skills to self-evaluate and mainstream evaluation within their organization. Fetterman outlines three main steps for conducting empowerment evaluation: (1) Develop and refine the "mission," (2) take stock and prioritize the program's activities, and (3) plan for the future.

Empowerment Evaluation involves three key principles

i. **Improvement:** The evaluation is designed to promote program improvement and to build the capacity of stakeholders to participate in and use evaluations effectively.

ii. **Participation**

iii. Stakeholders are actively involved in all stages of the evaluation process, and their input and feedback is valued and used.

iv. **Social Justice:** The evaluation is grounded in a social justice framework, which emphasizes the importance of promoting equity, inclusion, and empowerment.

Empowerment Evaluation is useful for evaluations that aim to promote social change and empower communities or organizations. It is often used in evaluations of community-based programs, where stakeholders have a vested interest in the program's success and are motivated to participate in the evaluation process. Empowerment Evaluation is also used in evaluations of programs that serve marginalized or underrepresented populations, where the goal is to build capacity and promote equity and social justice.

vii. Theory-Driven Evaluation

Huey Chen, PhD, is one of the main contributors to Theory-Driven Evaluation. His approach focuses on the theory of change and causal mechanisms underlying the program. Chen recognizes that programs exist in an open system, consisting of inputs, outputs, outcomes, and impacts. He suggests that evaluators should start by working with stakeholders to understand the assumptions and intended logic behind the program. A logic model can be used to illustrate the causal relationships between activities and outcomes. Chen offers many suggestions for constructing program theory models, such as action model (i.e. systematic plan for arranging staff, resources, settings to deliver services) and change model (i.e. set of descriptive assumptions about causal processes underlying intervention and outcome). Evaluators should consider using this approach when working with program implementers to produce valuable information for formative program improvement.

viii. Logic Model Theory

Logic Model Theory is an evaluation theory that emphasizes the importance of developing a clear and logical framework for program planning and evaluation. The theory emphasizes the need to clearly articulate the inputs, activities, outputs, outcomes, and impact of a program in a logical and coherent way, to facilitate program planning, implementation, and evaluation. In a logic model, the program's inputs are the resources that are available to the program, including funding, staff, and other resources. The activities are the program's interventions, or the actions taken to achieve the program's goals. The outputs are the direct products or services of the program, such as the number of participants served or the number of events held. The outcomes are the short-term and intermediate-term changes that occur as a result of the program, such as changes in knowledge, attitudes, or behaviors. The impact is the long-term change that occurs as a result of the program, such as improved health outcomes or reduced rates of crime. The logic model provides a visual representation of the program and the relationships between the program's components. It helps to clarify the program's goals and objectives, and to identify the inputs and activities that are most likely to lead to the desired outcomes and impact. The logic model can also be used to guide program implementation and to monitor and evaluate program performance. Logic Model Theory is useful for evaluations of complex programs or initiatives, where a clear and logical framework is necessary to guide program planning and evaluation. It is often used in program evaluations, policy evaluations, and organizational evaluations, and can be applied to both qualitative and quantitative data.

ix. Causal Loop Diagrams

Causal loop diagrams (CLDs) are a tool used in systems thinking and evaluation to visualize the complex causal relationships that exist between different components of a system. A causal loop diagram consists of a set of interconnected loops that represent the relationships between different components of a system, including the feedback loops that drive system behavior.

CLDs are useful for understanding the complex interactions that exist within a system and for identifying the key drivers of system behavior. They are often used in program evaluation to help program managers and evaluators understand the factors that contribute to program success or failure, and to identify potential areas for improvement. The process of developing a causal loop diagram typically involves several steps, including:

i. **Identifying the key components of the system:** This involves identifying the key elements of the system that are relevant to the program or intervention being evaluated.

ii. **Mapping the causal relationships between components:** This involves identifying the causal relationships between different components of the system and representing these relationships in the form of interconnected loops.

iii. **Identifying feedback loops:** This involves identifying the feedback loops that exist within the system and understanding how these feedback loops drive system behavior.

iv. **Analyzing the diagram:** This involves analyzing the diagram to identify the key drivers of system behavior and to identify potential areas for improvement.

Overall, causal loop diagrams are a powerful tool for understanding complex systems and for identifying the factors that contribute to program success or failure. By visualizing the causal relationships and feedback loops that exist within a system, program managers and evaluators can better understand the drivers of system behavior and make data-driven decisions to improve program effectiveness and impact.

x. Stock and flow diagrams

Stock and flow diagrams are a tool used in systems thinking and evaluation to represent the dynamic relationships that exist between different components of a system. Stock and flow diagrams are used to visualize the inflows and outflows of materials, energy, or other resources within a system over time. A stock and flow diagram consists of two types of components: stocks and flows. Stocks represent the accumulation of resources within a system, such as the amount of water in a reservoir, the number of people in a population, or the amount of money in a bank account. Flows represent the movement of resources within the system, such as the flow of water into or out of a reservoir, the flow of people into or out of a population, or the flow of money into or out of a bank account. Stock and flow diagrams are useful for understanding the behavior of complex systems over time and for identifying the key drivers of system behavior. They are often used in program evaluation to help program managers and evaluators understand the factors that contribute to program success or failure, and to identify potential areas for improvement.

The process of developing a stock and flow diagram typically involves several steps, including:

i. **Identifying the key components of the system:** This involves identifying the key stocks and flows within the system that are relevant to the program or intervention being evaluated.

ii. **Mapping the relationships between components:** This involves identifying the relationships between different stocks and flows within the system and representing these relationships in the form of a diagram.

iii. **Analyzing the diagram:** This involves analyzing the diagram to identify the key drivers of system behavior and to identify potential areas for improvement.

Overall, stock and flow diagrams are a powerful tool for understanding the dynamics of complex systems and for identifying the factors that contribute to program success or failure. By visualizing the inflows and outflows of resources within a system over time, program managers and evaluators can better understand the drivers of system behavior and make data-driven decisions to improve program effectiveness and impact.

xi. Concept map

A concept map is a tool used in evaluation and research to visually represent the relationships between different concepts or ideas. Concept maps are useful for organizing and synthesizing information, and for identifying the key relationships and themes that exist between different concepts. A concept map consists of a set of nodes, which represent different concepts or ideas, and a set of links or connectors, which represent the relationships between these concepts or ideas. The links can be directional or bidirectional and can represent different types of relationships, such as causal relationships, hierarchical relationships, or associative relationships. Concept maps are useful for a range of evaluation and research activities, including literature reviews, program planning and development, and data analysis. They can help to identify the key themes and relationships within a complex set of data or ideas and can provide a visual representation of these relationships that is easy to understand and communicate.

The process of developing a concept map typically involves several steps, including:

i. **Identifying the key concepts or ideas:** This involves identifying the key concepts or ideas that are relevant to the evaluation or research question being addressed.

ii. **Mapping the relationships between concepts:** This involves identifying the relationships between different concepts or ideas and representing these relationships in the form of a concept map.

iii. **Analyzing the concept map:** This involves analyzing the concept map to identify the key themes and relationships within the data or ideas being analyzed.

A concept map is a useful tool for organising and synthesizing information, as well as locating the core ideas and relationships included within complicated data or concepts. Concept maps are useful tools for assessors and researchers because they provide a visual representation of the links between various concepts or ideas. This helps both parties better comprehend and explain the results of their work.

xii. Network map

A network map is a tool used in evaluation and research to visually represent the relationships between different actors, organizations, or entities within a system. Network maps are useful for understanding the structure and dynamics of complex systems, and for identifying the key actors and relationships within these systems. A network map consists of a set of nodes, which represent different actors or entities, and a set of links or connectors, which represent the relationships between these actors or entities. The links can be directional or bidirectional and can represent different types of relationships, such as collaboration, communication, or influence. Network maps are useful for a range of evaluation and research activities, including stakeholder analysis, program planning and development, and policy analysis. They can help to identify the key actors and relationships within a system and can provide insights into the structure and dynamics of these systems. The process of developing a network map typically involves several steps, including:

i. **Identifying the key actors or entities:** This involves identifying the key actors or entities within the system that are relevant to the evaluation or research question being addressed.

ii. **Mapping the relationships between actors:** This involves identifying the relationships between different actors or entities and representing these relationships in the form of a network map.

iii. **Analyzing the network map:** This involves analyzing the network map to identify the key actors and relationships within the system, and to identify potential areas for improvement.

A network map is an effective method for gaining knowledge of the structure and dynamics of complex systems, as well as for locating the important players and links that are present within these systems. Network maps can assist evaluators and researchers in better understanding and communicating the findings of their work, as well as in identifying potential strategies for improving the efficiency of programs or systems. This is accomplished by visually representing the relationships that exist between various actors or entities.

xiii. Path Model

A path model is a statistical tool used in evaluation and research to analyze the relationships between multiple variables or factors. Path models are useful for understanding the complex relationships between different variables and for identifying the key drivers of program or system outcomes. A path model consists of a set of variables, which represent different factors that may influence program or system outcomes, and a set of arrows or paths, which represent the hypothesized relationships between these variables. Path models can be used to test hypotheses about the causal relationships between variables and to identify the key variables that are most strongly related to program or system outcomes. Path models are useful for a range of evaluation and research activities, including program evaluation, policy analysis, and impact evaluation. They can help to identify the key drivers of program or system outcomes and to identify potential areas for improvement.

The process of developing a path model typically involves several steps, including:

i. **Identifying the key variables:** This involves identifying the key variables that are relevant to the program or system being evaluated.

ii. **Specifying the hypothesized relationships:** This involves specifying the hypothesized relationships between the variables, based on theory or prior research.

iii. **Estimating the path coefficients:** This involves estimating the strength and direction of the relationships between variables using statistical methods.

iv. **Evaluating the model fit:** This involves evaluating the overall fit of the path model to the data, and making any necessary modifications to improve the fit.

Overall, path models are a powerful tool for understanding the complex relationships between different variables and for identifying the key drivers of program or system outcomes. By identifying the key variables that are most strongly related to program or system outcomes, path models can help evaluators and researchers to make data-driven decisions to improve program effectiveness and impact.

xiv. Realist Evaluation Theory

Realist Evaluation is an evaluation theory that focuses on understanding how interventions work in different contexts, and for whom, and why. The theory emphasizes the need to identify the underlying mechanisms that explain how and why a program or intervention works, and how these mechanisms interact with the context in which the program operates.

Realist Evaluation Theory is based on the assumption that programs are complex and are influenced by a variety of factors, including the context in which the program operates, the mechanisms that underlie the program, and the interactions between the program and its stakeholders. Realist Evaluation seeks to identify the context-mechanism-outcome configurations that explain how and why a program works or doesn't work.

Realist Evaluation involves several key steps

i. **Developing program theories:** The first step in Realist Evaluation is to develop program theories that explain how the program is expected to work, and under what conditions.

ii. **Testing program theories:** The second step is to test the program theories by collecting data on the program and analyzing it to identify the context-mechanism-outcome configurations.

iii. **Refining program theories:** The third step is to refine the program theories based on the findings of the evaluation, and to identify areas for improvement and further testing.

Realist Evaluation is useful for evaluations of complex programs or interventions, where a clear understanding of the mechanisms that underlie the program is necessary to identify the factors that contribute to the program's success or failure. It is often used in evaluations of health interventions, social programs, and education programs. Realist Evaluation can be applied to both qualitative and quantitative data, and can be used to inform program design, implementation, and evaluation.

xv. Nested and Hybrid Models

There are many different sorts of models, some of which have been shown in the sections that came before this one. In addition to this, there is also the potential of combining the various types of models. Each distinct kind of program theory has its own set of benefits, which, when combined with the others, may be of great use in the development and assessment of programs. In what follows, we are going to look at two different possibilities: nested models and hybrid models. Nested and hybrid models are two types of statistical models used in evaluation and research to analyze complex data and relationships between variables. A nested model is a type of hierarchical model in which lower-level units of analysis (such as individuals) are nested within higher-level units (such as organizations or communities). Nested models are useful for analyzing data with a hierarchical structure, and for accounting for the non-independence of observations within higher-level units.

- A hybrid model is a type of statistical model that combines elements of different types of models, such as linear regression, logistic regression, or path analysis. Hybrid models are useful for analyzing complex relationships between variables that cannot be adequately explained by a single type of model.

- Nested and hybrid models are useful for a range of evaluation and research activities, including program evaluation, impact evaluation, and policy analysis. They can help to identify the key drivers of program or system outcomes and to identify potential areas for improvement.

The process of developing a nested or hybrid model typically involves several steps, including:

i. **Identifying the key variables:** This involves identifying the key variables that are relevant to the program or system being evaluated.

ii. **Selecting the appropriate model**: This involves selecting the appropriate statistical model based on the structure of the data and the research question being addressed.

iii. **Estimating the model parameters:** This involves estimating the model parameters using statistical methods, such as maximum likelihood estimation.

iv. **Evaluating the model fit:** This involves evaluating the overall fit of the model to the data, and making any necessary modifications to improve the fit.

Overall, nested and hybrid models are powerful tools for analyzing complex data and relationships between variables, and for identifying the key drivers of program or system outcomes. By identifying the key variables that are most strongly related to program or system outcomes, nested and hybrid models can help evaluators and researchers to make data-driven decisions to improve program effectiveness and impact.

Integration between theory and practice of evaluation

a) Evaluation forums

b) Workshops

c) Conferences

d) Apprenticeship/ internship.

Block 2: Conducting Evaluation

Unit 1

How to Conduct Evaluation

Ten Steps in programme evaluation

i. **Identify and describe programme you want to evaluate:** The first step of the evaluation process is to identify the object or programme of evaluation. For this purpose, a short written description of the object or programme to be evaluated must be drafted. This description must be formulated in a way that is also comprehensible to outsiders. Relevant items to be included in this description are, for example, the name of the object or programme of evaluation, the expert responsible for it, the employees taking part in it, the goals, scientific references, the contents or topics, the elements, methods, scope, available resources, key figures, past developments, etc.

ii. **Identify the phase of the programme**

- Design
- Start-up
- Ongoing
- Wrap-up
- Follow-up

Type of evaluation study needed

- Needs assessment (Rational and Empirical methods)
- Baseline survey
- Formative, Ongoing, Process, Summative evaluation etc.

iii. **Assess the feasibility of implementing an evaluation:** The evaluation team conducted a feasibility assessment and concludes that it is feasible to implement the evaluation process. At this point, the first deliverable, a statement of work for conducting an evaluation of the program has been prepared. It consisted of information about the program to be evaluated, the purpose of the evaluation, the scope of work and timeline, resources to be used by the evaluation team and committed by the client, assumptions and risks.

iv. **Identify and consult key stakeholders:** It is important that the evaluation team identify and consult with three types of stakeholders for the program, in order to not only better understand their needs and clearly identify the purpose of the evaluation, but also estimate appropriate sources of data to be used later in the evaluation.
   - Upstream stakeholders
   - Direct impactees/Addressees' or 'Intended users
   - Indirect impactees

v. **Identify approaches to data collection**
   - Quantitative method
   - Qualitative method
   - Mixed method

vi. **Select data collection techniques**
   - Survey interviews
   - Questionnaires with different types

vii. **Identify population and select sample**
   - Sampling for evaluation
   - Sample size
   - Errors
   - Sampling techniques

viii. **Collect, analyse and interpret data:** Carefully planned data collection must occurs in the evaluation process. Evaluations do not take place under laboratory conditions, but are applied social science endeavours affected by a dynamic learning and working environment, which in turn is influenced by the evaluation itself. In such a setting, data collection should be integrated as smoothly as possible into the 'everyday activities' of the object of evaluation. Data is analyzed by two ways i.e.
   - Qualitative evaluation data analysis
   - Quantitative evaluation data analysis

ix. **Communicate findings:** The reporting is carried out in a way that enables the addressees of the evaluation to utilize the evaluation findings. Ideally, reporting should be planned at an early stage of the evaluation process. The following points must be clarified: who will be the addressees? What informational interests do they have with regard to the reporting process? When will findings be reported? Who will be responsible for preparing the interim and final reports? In what format will reports be submitted? Following points should be taken into consideration before communicating the findings,

- Reporting plan
- Evaluation report types
- Reporting results
- Reporting tips
- Reporting negative findings

x. **Apply and use findings:** Evaluations are only worthwhile if they are used. For this reason, systematically preparation and encouraging the utilization of the evaluation is just as much a part of the ninth step of the evaluation process as the actual utilization of the evaluation itself. Evaluation helps in taking decision on the following;

- Programme continuation
- Discontinuation,
- Improve on-going programme,
- Plan future programmes
- Inform programme stakeholders

Evaluating the Evaluation

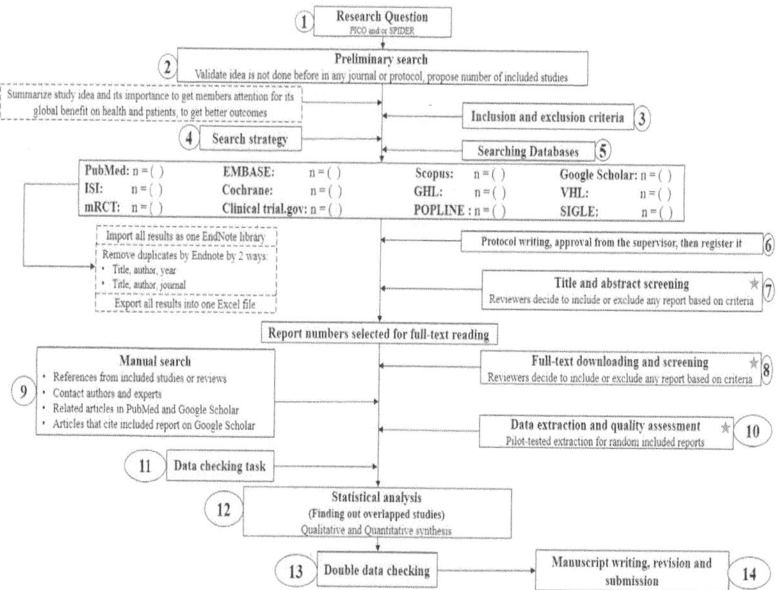

Detailed flow diagram guideline for systematic review and meta-analysis steps
(*Source:* Tawfik *et al.*, 2019)

Block 3: Programme Management Techniques

Unit 1

SWOT Analysis and Bar Charts

SWOT Analysis – Concept, origin and evolution

While working on a research project at the Stanford University sometime around ***1960s to 1970s, Albert Humphrey*** developed an analytical tool to evaluate the strategic plans and find out why corporate planning failed. He coined this technique as **SOFT** analysis where –

- **S** stood for what things are Satisfactory at present,
- **O** denoted what Opportunities can be explored in the future,
- **F** meant the Faults in the present and
- **T** signified the Threats that could surface in the future.

While the majority agrees that SOFT is the predecessor of SWOT, some people believe that the concept of SWOT analysis emerged separately and has nothing to do with SOFT.

How SWOT evolved from SOFT?

The first mention of the *term SWOT* can be traced back to the Long Range Planning seminar held in *Zurich in 1964*. In this seminar Urick and Orr proposed the concept of SWOT analysis which was derived from SOFT by replacing the F for faults with W for Weaknesses. With its initial promotion in the Britain, the concept soon gained recognition among strategic planners and management consultants the world over. The new interpretation of SWOT, the way it is used these days, differs slightly from that of SOFT.

- The **S** in SWOT now represents the strength – which give the business an advantageous edge over the other similar businesses.
- **W** stands for weaknesses and these weaknesses can be understood as the comparative disadvantages that a business faces as against other similar businesses.
- **O** represents opportunities – the external possibilities that can be explored or utilized to boost the business's growth.
- **T** signifies threats – the external or environmental factors which can cause problems for the business.

Emerging of the SWOT matrix

Another important development in the history of the SWOT analysis was the development of the SWOT matrix. In 1982, Dr Heinz Weihrich proposed the use of a *2x2 matrix* for carrying out a SWOT analysis. This matrix was initially popular as TOWS matrix, which is another name used for referring to a SWOT matrix even today.Ever since the 1980s, the SWOT interests management professionals and forms an integral part of strategic planning mechanism. Looking at history, one can see that a lot many similar concepts were introduced during various researches, but none of them survived for too long. In fact, some of the tools were very similar to SWOT. History is proof enough that SWOT is by far the best and the most widely used strategic planning tool.

SWOT or SWOT Matrix As a Programme Management Tool

SWOT analysis is a strategic planning tool that comes in handy for identifying and analyzing the objectives and plans of a business or project in terms of strengths, weaknesses, opportunitics and threats. The best way to present these findings is using a 2X2 SWOT matrix.

Components of a SWOT Matrix

As mentioned earlier the SWOT analysis requires four types of information:

- **Strengths:** Under strengths the user can enlist all characteristics or factors that the business or the project has a comparative advantage in. Any positive trait that gives the project an edge over the competing projects goes into the cell reserved for the strengths. In order to find out the strengths make a list of the things that the project or business is good at doing.
- **Weaknesses:** Weaknesses are factors that take away the competitive edge from the business or project and make it fall short when compared against its competitors. In other words weaknesses are the internal negative points which can create problems in achieving the goals and objectives. Weakness can be easily found out by analyzing the things that are lacking and that can pull down the performance.
- **Opportunities:** Opportunities are things that the business or the project has not explored and which can be potentially rewarding. The best way to find out the opportunities is to look at the competitors and see what new avenues they are exploring.
- **Threats:** External factors which are outside the control of the business but can have a negative effect on it can be classified as threats. Threats can be found out by scanning the operating environment for unforeseen developments that can prove detrimental for the business or project.

Creating a SWOT Matrix

These four components of a SWOT matrix can be grouped in two ways – on the basis of their scope and on the basis of their nature.

On the basis of the scope the four components of SWOT can be classified as:

- **Internal Factors:** The strengths and weaknesses are the internal factors.
- **External Factors:** Opportunities and threats fall under the external factors.

On the basis of nature, these components can be again grouped under two categories:

- **Positive Factors:** Strengths and opportunities are both good and are thus classified as positive factors.
- **Negative Factors:** Weaknesses and threats are the negative factors which have a bad influence on the business or the project.

These groupings can be presented easily in the form of the two axis of the matrix, as can be seen in the SWOT matrix template. A SWOT matrix is indeed a useful strategic planning tool but its effectiveness depends on how properly the SWOT analysis has been conducted. It's important to write down only precise and workable statements ito the matrix. Prioritizing the lists is equally important.

Conducting a SWOT Analysis

SWOT analysis is normally led by a facilitator who has considerable responsibility for executing the company's strategy. The following steps describe how to conduct a SWOT analysis:

1. Invite representatives from finance, operation, marketing, and any other key players.
2. Clearly explain the purpose for conducting the SWOT analysis and how you will do it.
3. Hang out four large flipcharts, one for each of the four SWOT categories.
4. Brainstorm the strengths and weaknesses within your business, and the opportunities and threats present in your environment.
5. Record inputs and ideas from the brainstorming session on the flipcharts.
6. Clarify content and ensure the appropriateness and completeness of the recorded information.
7. Share and communicate the outcomes to relevant stakeholders.
8. Take actions and assign responsibilities to maximize strengths and opportunities and minimize weaknesses and threats.

Example

The following example uses a four-field matrix to present the outcome of a SWOT analysis for an organization. Note that it is normal for any business to have weaknesses and potential threats. Attempts should be made to convert weaknesses into strengths and threats into opportunities.

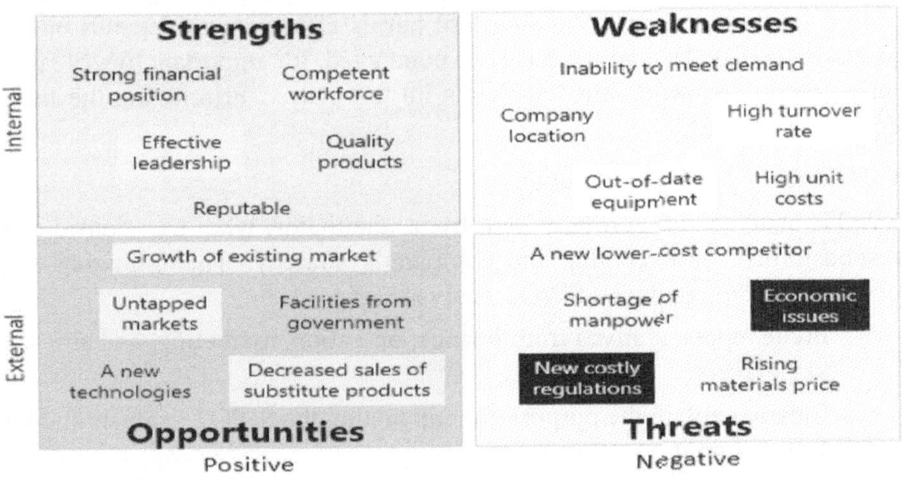

Personal SWOT Analysis

Although SWOT analysis was originally made for businesses, it can be used with equal success to identify and understand a person's strengths, weaknesses, opportunities and threats. The analysis can help you to better understand many things about yourself and your external environment. You can then apply personal development strategies to turn weaknesses into strength, take advantages of strengths and opportunities, and minimize or eliminate weaknesses and threats.

Example

The following is an example that uses SWOT analysis to identify and assess a person's strengths, weaknesses, opportunities and threats.

Common Questions in SWOT Analysis

Implementation of SWOT is achievable by asking questions such as the following:

Strengths

What are the organization's advantages?

What can you do better than others?

What unique or lowest-cost services can you provide patients?

What do patients in your market see as your organization's strength?

Weaknesses

Upon what factors could the organization improve?

What are patients in your market likely to see as your organization's weakness?

What lack of services loses your organization patients?

Opportunities

What good opportunities are available to your organization?

What are the new and exciting trends your organization can try?

What new changes to governmental regulation/policy can benefit your organization?

Threats

What problems does your organization face?

Of what are your organization's competitors taking advantage?

Do evolving technologies and new services threatening your organization's position in the minds of patients?

Does your facility have cash-flow problems?

Could any of your weaknesses threaten quality patient care?

Advantages of SWOT

Conducting a SWOT analysis can help when you're at the starting point of a project. It can also help project managers optimize planning and efficiency by staying aware of potential threats and weaknesses. Running a SWOT analysis can also benefit an organization in several ways:

i. Identify strengths and weaknesses associated with your project/programme/enterprise so that you can focus on areas needing improvement
ii. Find potential threats that may arise in order to make well-informed decisions
iii. Improve collaboration by bringing people together to discuss how to approach a project
iv. Simplify the brainstorming process by establishing a shared understanding of a project's current state
v. Inform the strategic planning process by creating a framework for analyzing a situation and making future plans

Disadvantages of SWOT

- Some SWOT analysis users oversimplify the amount of data used for decisions – it's easy to use insufficient data.
- The risk of capturing too much data may lead to 'paralysis by analysis'.
- The data used may be based on assumptions that later prove to be unfounded.
- Access to quality internal data sources can be time consuming and politically difficult (especially in more complex organisations – parent company, etc).
- It lacks detailed structure, so key elements may get missed.

- The pace of change makes it increasingly difficult to anticipate developments that may affect an organisation in the future.
- To be effective, the process needs to be repeated on a regular basis.

A SWOT analysis can be used for

- Workshop sessions.
- Generating ideas and solutions.
- Problem solving.
- Planning.
- Strategic planning (with PESTLE).
- Product evaluation.
- Competitor evaluation (with Porter's five forces).
- Personal development planning.
- Decision making (with Lewin's force field analysis).

Bar Charts

A **bar chart** or **bar graph** is a chart or graph that presents categorical data with rectangular bars with heights or lengths proportional to the values that they represent. The bars can be plotted vertically or horizontally. A vertical bar chart is sometimes called a **column chart**.

A bar graph shows comparisons among discrete categories. One axis of the chart shows the specific categories being compared, and the other axis represents a measured value. Some bar graphs present bars clustered in groups of more than one, showing the values of more than one measured variable.

History

William Playfair (1759-1824) considered to have invented the bar chart and the *Exports and Imports of Scotland to and from different parts for one Year from Christmas 1780 to Christmas 1781* graph from his *The Commercial and Political Atlas* to be the first bar chart in history. Diagrams of the velocity of a constantly accelerating object against time published in *The Latitude of Forms* (attributed to Jacobus de Sancto Martino or, perhaps, to Nicole Oresme) about 300 years before can be interpreted as "proto bar charts"

Types

i. **Grouped (clustered) bar charts:** for each categorical group there are two or more bars color-coded to represent a particular grouping. For example, a business owner with two stores might make a grouped bar chart with different colored bars to represent each store: the horizontal axis would show the months of the year and the vertical axis would show revenue.

ii. **Stacked bar chart:** stacks bars on top of each other so that the height of the resulting stack shows the combined result. Stacked bar charts are not suited to data sets having both positive and negative values.

Advantages

1. **Easy to read and interpret:** Bar charts are easy to read and interpret, even for people without a background in statistics or data visualization. The bars make it easy to compare values and see trends, making it a useful tool for communicating information to a wide range of audiences.
2. **Can handle large amounts of data:** Bar charts can handle large amounts of data and still provide a clear representation of the information. The bars can be made narrow or wide to fit a large number of categories or data points, and the use of color or patterns can make it easier to distinguish between them.
3. **Customizable:** Bar charts can be customized to suit the needs of the user. For example, the color, width, and height of the bars can be adjusted to make the chart more visually appealing, and labels and annotations can be added to provide additional information.
4. **Useful for comparing values:** Bar charts are particularly useful for comparing values between categories or data points. They allow for quick identification of differences and similarities, making it easy to draw conclusions and make decisions.

Limitations

1. **Limited use for continuous data:** Bar charts are not useful for displaying continuous data, such as temperature or time. For continuous data, a line chart or scatter plot may be more appropriate.
2. **Limited use for small sample sizes:** Bar charts may not be useful for displaying small sample sizes, as the bars may not accurately represent the data. In such cases, a histogram or box plot may be more appropriate.
3. **May be misleading:** Bar charts can be misleading if the scale is not appropriate or if the data is presented in a way that is designed to mislead the viewer. For example, if the y-axis is truncated, the differences between the bars may appear larger than they actually are.
4. **Limited scope for multivariate data:** Bar charts can only display one or two variables at a time, making them less useful for displaying multivariate data. In such cases, a scatter plot or heat map may be more appropriate.

Gantt Charts

A Gantt chart is a type of bar chart that illustrates a project schedule. This chart lists the tasks to be performed on the vertical axis, and time intervals on the horizontal axis. The width of the horizontal bars in the graph shows the duration of each activity. Gantt charts illustrate the start and finish dates of the terminal elements and summary elements of a project. Terminal elements and summary elements constitute the work breakdown structure of the project. Modern Gantt charts also show the dependency (i.e., precedence network) relationships between activities. Gantt charts can be used to show current schedule status using percent-complete shadings and a vertical «TODAY» line. Gantt charts are sometimes equated with bar charts. Gantt charts are usually created initially using an *early start time approach*, where each task is scheduled to start immediately when its prerequisites are complete. This method maximizes the float time available for all tasks. According to Wikipedia, "A Gantt chart is a type of bar chart that illustrates a project schedule and shows the dependency relationships between activities and current schedule status."

History

- The first known tool of this type was developed in 1896 by **Karol Adamiecki,** who called it a ***harmonogram***. Adamiecki, however, published his chart only in Russian and Polish which limited both its adoption and recognition of his authorship.
- In 1912, Hermann Schürch published what could be considered Gantt charts while discussing a construction project. Charts of the type published by Schürch appear to have been in common use in Germany at the time; however, the prior development leading to Schürch's work is unclear. Unlike later Gantt charts, *Schürch's charts did not display interdependencies,* leaving them to be inferred by the reader. These were also static representations of a planned schedule.
- The chart is named after Henry Gantt (1861–1919), who designed his chart around the years **1910–1915.** Gantt originally created his tool for systematic, routine operations. He designed this visualization tool to more easily measure productivity levels of employees and gauge which employees were under- or over-performing. Gantt also frequently included graphics and other visual indicators in his charts to track performance.
- One of the first major applications of Gantt charts was by the United States during World War I, at the instigation of General William Crozier.

- Gantt's collaborator Walter Polakov introduced Gantt charts to the Soviet Union in 1929 when he was working for the Supreme Soviet of the National Economy. They were used in developing the First Five Year Plan, supplying Russian translations to explain their use.
- In the 1980s, personal computers allowed widespread creation of complex and elaborate Gantt charts. The first desktop applications were intended mainly for project managers and project schedulers. With the advent of the Internet and increased collaboration over networks at the end of the 1990s, Gantt charts became a common feature of web-based applications, including collaborative groupware.
- By 2012, almost all Gantt charts were made by software which can easily adjust to schedule changes.
- In 1999, Gantt charts were identified as "one of the most widely used management tools ***for project scheduling and control***".

Example

In the following tables there are seven tasks, labeled *a* through *g*. Some tasks can be done concurrently (*a* and *b*) while others cannot be done until their predecessor task is complete (*c* and *d* cannot begin until *a* is complete). Additionally, each task has three time estimates: the optimistic time estimate (*O*), the most likely or normal time estimate (*M*), and the pessimistic time estimate (*P*). The expected time (T_E) is estimated using the beta probability distribution for the time estimates, using the formula $(O + 4M + P) \div 6$.

Activity	Predecessor	Time estimates (in days)			Expected time (TE)
		Opt. (0)	Normal (M)	Pess. (P)	
a	-	2	4	6	4.00
b	-	3	5	9	5.33
c	a	4	5	7	5.17
d	a	4	6	10	6.33
e	b, c	4	5	7	5.17
f	d	3	4	8	4.50
g	e	3	5	8	5.17

Once this step is complete, one can draw a Gantt chart or a network diagram.

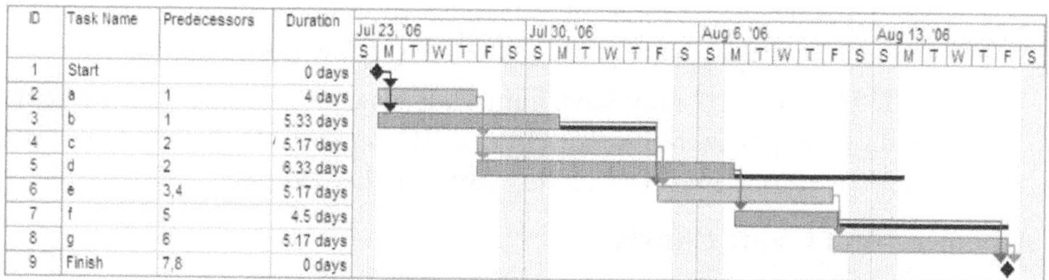

ID	Task Name	Predecessors	Duration
1	Start		0 days
2	a	1	4 days
3	b	1	5.33 days
4	c	2	5.17 days
5	d	2	6.33 days
6	e	3,4	5.17 days
7	f	5	4.5 days
8	g	6	5.17 days
9	Finish	7,8	0 days

A Gantt chart created using Microsoft Project. Note (1) the critical path is in red, (2) the slack is the black lines connected to non-critical activities, (3) since Saturday and Sunday are not work days and are thus excluded from the schedule, some bars on the Gantt chart are longer if they cut through a weekend.

Characteristics of Gantt Charts

i. ***DUE DATES***: The dates are one of the most essential aspects of a Gantt chart since they show project managers not only when the project will start and end, but also when each job will take place. These are shown at the top of the graph.

ii. ***TASKS:*** Major projects always seem to have a lot of sub-tasks. A Gantt chart assists project managers in keeping track of all sub-tasks in a project so that nothing is overlooked or delayed. The tasks are displayed on the left side of the page.

iii. ***MILESTONES:*** Milestones are tasks that are critical to the completion and success of a project. Unlike the minor things that must also be completed, completing a milestone provides a sense of accomplishment and progress. At the conclusion of each block on a Gantt chart, milestones are represented by different shapes or icons.

iv. ***BARS:*** Bars are used to represent the time frame in which each task should be performed after the subtasks have been stated. This ensures that each sub-task is finished on time, ensuring that the overall project is completed on time.

v. ***TASKBARS:*** While many sub-tasks can be accomplished pretty quickly, there will be occasions when you want to know how your project is progressing at a glance. The taskbars are shaded to represent the portion of each task that has previously been performed, indicating progress.

vi. ***DEPENDENCIES:*** In a project, there are some tasks and subtasks that are dependent on one another for success. For instance, a task must be completed before another task can begin or terminate. On a Gantt chart, task dependencies represent this type of relationship. Small arrows between the taskbars are generally used to show these relationships.

vii. ***TASK ID:*** You probably have numerous tasks going on at the same time in today's hyper work world. The task ID is included on the Gantt chart to help everyone involved readily identify the task you're discussing.

Advantages of Gantt Charts

i. It is to represent the Project schedules and Activities
ii. Easy to represent Tasks, Sub-tasks, Milestones and Projects Visually on a Graph
iii. Clear visibility of Dates and Time Frames
iv. It helps to see the Plans by Day, Week, Month, Quarter and Year
v. Helps to effectively manage the Team
vi. And it helps in efficient Time Management
vii. Easy to group all sub tasks under a main task
viii. Also, we can see the Team Members and their responsible tasks
ix. Easy to Check the Project Status
x. We can See the Completed % of Tasks
xi. Tasks in Progress and Pending work is clearly visible on Stacked Bars
xii. Helps Managers to easily coordinate with the teams
xiii. Gantt chart is good tool for presenting in Team Meetings

Limitations of Gantt Charts

i. Require more efforts for Creating and Managing the Chart
ii. Updating a Chart is Very Time Consuming
iii. All Tasks are not visible in a single view of a Gantt
iv. Need to scroll and Click additional buttons to view remaining items
v. Stacks represents only the time and not the hours of the work
vi. Not easy to re align the tasks from on section to another
vii. Not easy to calculate the aggregates

Milestone Charts

A milestone chart is a visual project planning tool that uses milestones to divide a project plan into major phases. Each milestone indicate the achievement which brings closer to complete the project—in this way, milestones provide a sense of accomplishment and show the team how the work they're doing contributes to the overarching project objective. Due to its simplicity, it's used when project managers or sponsors need to share an overview of the project schedule with stakeholders or team members without going over every project task.

Importance of a Milestone Chart

A Milestone Chart is a crucial project management tool that helps teams stay on track by visually representing key goals and deadlines. Here's why it's important:

1. **Clear Project Overview**
   - Provides a **high-level summary** of the project timeline.
   - Helps stakeholders quickly understand progress and key phases.
2. **Improved Planning & Scheduling**
   - Defines critical deadlines, preventing last-minute rushes.
   - Helps in **allocating resources** efficiently based on project needs.
3. **Enhances Team Accountability**
   - Assigns clear responsibilities for each milestone.
   - Encourages team members to stay on schedule.
4. **Early Problem Detection**
   - Highlights **bottlenecks or delays** in project progress.
   - Allows for proactive problem-solving before issues escalate.
5. **Effective Communication**
   - Acts as a **visual reference** for project updates.
   - Keeps all stakeholders aligned, including clients, managers, and team members.
6. **Boosts Motivation & Productivity**
   - Celebrating milestone achievements can keep the team engaged.
   - Provides a sense of progress and accomplishment.
7. **Helps in Decision-Making**
   - Enables project managers to make informed adjustments.
   - Supports data-driven decisions by tracking past milestones.

Steps to Create a Milestone Chart

1. **Define the Project Scope**
   - Identify the overall goal and major deliverables.
   - Break the project into phases if necessary.

2. **Identify Key Milestones**
   - Milestones are significant events like project kick-off, approvals, completion of phases, or major deliverables.
   - Ensure each milestone has a clear purpose.
3. **Set Target Dates**
   - Assign realistic deadlines for each milestone.
   - Consider dependencies and resource availability.
4. **Choose a Format for the Chart**
   - Common formats include Gantt charts, timeline diagrams, or simple tabular formats.
   - Software like Microsoft Project, Excel, or online tools like Trello and Asana can help.
5. **Plot the Milestones on a Timeline**
   - Use a horizontal or vertical timeline to visualize the progression.
   - Ensure each milestone is labeled with a date and description.
6. **Assign Responsibilities**
   - Identify team members responsible for achieving each milestone.
   - Communicate expectations clearly.
7. **Monitor and Update Regularly**
   - Track progress and update the chart as needed.
   - Adjust dates or add new milestones if necessary.

Benefits of using a milestone chart

- **Clear project overview:** A milestone chart provides a high-level overview of the project. It highlights key dates and milestones in an easy-to-understand way.
- **Helps to track progress:** A milestone chart helps the team to monitor progress and identify potential delays or issues that may impact the timeline. Steps can then be taken in advance.
- **Manages expectations:** A milestone chart sets and manages expectations among stakeholders and team members. Everyone will have a clear idea of what is to be accomplished and how much remains to be done.
- **Facilitates communication and collaboration:** Because it is clear and concise, a milestone chart can clearly communicate project progress to stakeholders, clients, and team members. It also enables a collaborative environment by providing a shared understanding of goals, timelines, and expectations.

- **Improves decision-making:** A milestone chart can help project managers make decisions about resource allocation, project priorities, and risks.
- **Promotes accountability:** When milestones are linked to relevant team members on the chart, it creates a sense of accountability. All members get a clear picture of responsibilities and deadlines.

Unit 2

Networks

Networks – Introduction, origin and widely used networks

Network is a technique used for planning and scheduling of large projects in the fields of construction, maintenance, fabrication, purchasing, computer system instantiation, research and development planning etc. There is multitude of operations research situations that can be modelled and solved as network. Some recent surveys reports that as much as 70% of the real-world mathematical programming problems can be represented by network related models. Network analysis is known by many names PERT (Programme Evaluation and Review Technique), CPM (Critical Path Method), PEP (Programme Evaluation Procedure), LCES (Least Cost Estimating and Scheduling), SCANS (Scheduling and Control by Automated Network System), etc.

It is a graphical representation of logical and sequentially connected activities and events of a project. Network is also called arrow diagram. PERT (Programme Evolution Review Technique) and (Critical Path Method) are the two most widely applied techniques.

Programme Evaluation and Review Technique (PERT)

PERT was developed primarily to simplify the ***planning and scheduling*** of large and complex projects. The US Navy Special Projects Office, The Evaluation Office of the Lockheed Missile Systems Division, and the Operations Research Department of Booz-Allen-Hamilton developed PERT back in **1957** during the production of the UGM-27 Polaris missile of the US Navy. It found applications all over industry. An early example is when it was used for the 1968 Winter Olympics in Grenoble which applied PERT from 1965 until the opening of the 1968 Games. This project model was the first of its kind, a revival for scientific management, founded by Frederick Taylor (Taylorism) and later refined by Henry Ford (Fordism).

Initially PERT stood for ***Program Evaluation Research Task,*** but by **1959** was renamed. It had been made public in 1958 in two publications of the U.S. Department of the Navy, entitled ***Program Evaluation Research Task, Summary Report, Phase 1*** **and** ***Phase 2***. In a 1959 article in *The American*

Statistician the main Willard Fazar, Head of the Program Evaluation Branch, Special Projects Office, U.S. Navy, gave a detailed description of the main concepts of the PERT.

Project Evaluation and Review Technique (PERT) is a procedure through which activities of a project are represented in *its appropriate sequence and timing*. It is a *scheduling technique* used to *schedule, organize and integrate tasks* within a project. PERT is basically a mechanism for management *planning and control* which provides blueprint for a particular project. All of the primary elements or events of a project have been finally identified by the PERT. In this technique, a PERT Chart is made which represent a schedule for all the specified tasks in the project. The reporting levels of the tasks or events in the PERT Charts is somewhat same as defined in the work breakdown structure (WBS).

Characteristics of PERT

The main characteristics of PERT are as following:

i. It serves as a base for obtaining the important facts for implementing the decision-making.

ii. It forms the basis for all the planning activities.

iii. PERT helps management in deciding the best possible resource utilization method.

iv. PERT take advantage by using time network analysis technique.

v. PERT presents the structure for reporting information.

vi. It helps the management in identifying the essential elements for the completion of the project within time.

Advantages of PERT

It has the following advantages

1. Estimation of completion time of project is given by the PERT.
2. It supports the identification of the activities with slack time.
3. The start and dates of the activities of a specific project is determined.
4. It helps project manager in identifying the critical path activities.
5. PERT makes well organized diagram for the representation of large amount of data.
6. PERT chart explicitly defines and makes visible dependencies (precedence relationships) between the work breakdown structure (commonly WBS) elements.

7. PERT facilitates identification of the critical path and makes this visible.
8. PERT facilitates identification of early start, late start, and slack for each activity.
9. PERT provides for potentially reduced project duration due to better understanding of dependencies leading to improved overlapping of activities and tasks where feasible.
10. The large amount of project data can be organized and presented in diagram for use in decision making.
11. PERT can provide a probability of completing before a given time.

Disadvantages of PERT

It has the following disadvantages

1. The complexity of PERT is more which leads to the problem in implementation.
2. The estimation of activity time are subjective in PERT which is a major disadvantage.
3. Maintenance of PERT is also expensive and complex.
4. The actual distribution of may be different from the PERT beta distribution which causes wrong assumptions.
5. It under estimates the expected project completion time as there is chances that other paths can become the critical path if their related activities are deferred.
6. PERT is not easily scalable for smaller projects.
7. The network charts tend to be large and unwieldy, requiring several pages to print and requiring specially-sized paper.
8. The lack of a timeframe on most PERT charts makes it harder to show status, although colours can help, *e.g.*, specific colour for completed nodes.

Critical Path Method (CPM)

The CPM is a project-modeling technique developed in the ***late 1950s by Morgan R. Walker of DuPont and James E. Kelley Jr. of Remington Rand.*** Kelley and Walker related their memories of the development of CPM in 1989.Kelley attributed the term "critical path" to the developers of the PERT, which was developed at about the same time by ***Booz Allen Hamilton and the U.S. Navy.*** The precursors of what came to be known as critical path were developed and put into practice by DuPont between 1940 and 1943 and contributed to the success of the Manhattan Project.

Critical path analysis is commonly used with all forms of projects, including ***construction, aerospace and defense, software development, research projects, product development, engineering, and plant maintenance, among others***. Any project with interdependent activities can apply this method of mathematical analysis. CPM was used for the first time in 1966 for the major skyscraper development of constructing the former World Trade Center Twin Towers in New York City. Although the original CPM program and approach is no longer used,the term is generally applied to any approach used to analyze a project network logic diagram.

Components

The essential technique for using CPM is to construct a model of the project that includes:

1. A list of all activities required to complete the project (typically categorized within a work breakdown structure)
2. The time (duration) that each activity will take to complete
3. The dependencies between the activities
4. Logical end points such as milestones or deliverable items

Difference between PERT and CPM

Sl. No.	PERT	CPM
1.	PERT is that technique of project management which is used to manage uncertain (i.e., time is not known) activities of any project.	CPM is that technique of project management which is used to manage only certain (i.e., time is known) activities of any project.
2.	It is event oriented technique which means that network is constructed on the basis of event.	It is activity oriented technique which means that network is constructed on the basis of activities.
3.	It is a probability model.	It is a deterministic model.
4.	It majorly focuses on time as meeting time target or estimation of percent completion is more important.	It majorly focuses on Time-cost trade off as minimizing cost is more important.
5.	It is appropriate for high precision time estimation.	It is appropriate for reasonable time estimation.
6.	It has Non-repetitive nature of job.	It has repetitive nature of job.
7.	There is no chance of crashing as there is no certainty of time.	There may be crashing because of certain time boundation.
8.	It doesn't use any dummy activities.	It uses dummy activities for representing sequence of activities.
9.	It is suitable for projects which required research and development.	It is suitable for construction projects.

Networks Terminology

Events and activities

In a PERT diagram, the main building block is the *event,* with connections to its known predecessor events and successor events.

- ***PERT event*:** a point that marks the start or completion of one or more activities. It consumes no time and uses no resources. When it marks the completion of one or more activities, it is not "reached" (does not occur) until *all* of the activities leading to that event have been completed.
- ***Predecessor event*:** an event that immediately precedes some other event without any other events intervening. An event can have multiple predecessor events and can be the predecessor of multiple events.
- ***Successor event*:** an event that immediately follows some other event without any other intervening events. An event can have multiple successor events and can be the successor of multiple events.
- **Merge Events:** When more than one activity comes and joins an event. Such event is known as merge events.
- **Burst event:** When more than one activity leaves an events, such event is known as burst event.
- **Activity Duration:** This can be defined as time taken to complete a given activity.

Besides events, PERT also knows activities and sub-activities:

- ***PERT activity*:** the actual performance of a task which consumes time and requires resources (such as labor, materials, space, machinery). It can be understood as representing the time, effort, and resources required to move from one event to another. A PERT activity cannot be performed until the predecessor event has occurred.
- ***PERT sub-activity*:** a PERT activity can be further decomposed into a set of sub-activities. For example, activity A1 can be decomposed into A1.1, A1.2 and A1.3. Sub-activities have all the properties of activities; in particular, a sub-activity has predecessor or successor events just like an activity. A sub-activity can be decomposed again into finer-grained sub-activities.

Time

PERT has defined four types of time required to accomplish an activity:

- ***Optimistic time*:** the minimum possible time required to accomplish an activity (o) or a path (O), assuming everything proceeds better than is normally expected

- ***Pessimistic time***: the maximum possible time required to accomplish an activity (p) or a path (P), assuming everything goes wrong (but excluding major catastrophes).
- ***Most likely time***: the best estimate of the time required to accomplish an activity (m) or a path (M), assuming everything proceeds as normal.
- ***Expected time***: the best estimate of the time required to accomplish an activity (te) or a path (TE), accounting for the fact that things don't always proceed as normal (the implication being that the expected time is the average time the task would require if the task were repeated on a number of occasions over an extended period of time).
- ***Standard deviation of time***: the variability of the time for accomplishing an activity (σ_{te}) or a path (σ_{TE}).

Management tools

PERT supplies a number of tools for management with determination of concepts, such as:

- ***Float** or **slack*** is a measure of the excess time and resources available to complete a task. It is the amount of time that a project task can be delayed without causing a delay in any subsequent tasks (*free float*) or the whole project (*total float*).
  i. Positive slack would indicate *ahead of schedule*
  ii. Negative slack would indicate *behind schedule*
  iii. Zero slack would indicate *on schedule*
- ***Critical path***: the longest possible continuous pathway taken from the initial event to the terminal event. It determines the total calendar time required for the project; and, therefore, any time delays along the critical path will delay the reaching of the terminal event by at least the same amount.
- ***Critical activity***: An activity that has total float equal to zero. An activity with zero free float is not necessarily on the critical path since its path may not be the longest.
- ***Lead** time*: the time by which a *predecessor event* must be completed in order to allow sufficient time for the activities that must elapse before a specific PERT event reaches completion.
- ***Lag time***: the earliest time by which a *successor event* can follow a specific PERT event.

- ***Fast tracking***: performing more critical activities in parallel
- ***Crashing critical path***: Shortening duration of critical activities

Early Start Time-EST

Early Start Time or **EST** is the earliest time that activity can start. An activity near the end of path or much later will only start if all the previous activities in the path also started early. If any one of the previous activities slips then it will push this activity out.

Early Finish Time-EFT

Early Finish Time or **EFT** is the earliest time that activity can finish. It is the date that an activity can finish if all previous activities started early and none of the activities slipped. If project manager and team knows about early start and early finish dates of all the activities (tasks) then they will know how much freedom they have to move the start dates without causing any problem to the project.

Late Start Time-LST

Late Start Time or **LST** is the latest time an activity can start. If an activity is on the path which is much shorter compare to critical path, then it can start much late without any delay to the project. But these delay cause problem if other activities on the path slip.

Late Finish Time-LFT

Late Finish Time or **LFT** is the latest time that an activity can finish. If an activity is on a shorter path and all of the other activities on same path start as well as finish early, then it can finish very late without causing project to be late. Knowing late start and late finish will help you see how much it can be played around in the schedule. An activity having large LST and LFT means there are more options available.

Tips for calculation of EST, EFT during forward pass

- EST of first activity = 1
- EST of all other activites = EFT (of previous activity) + 1
- If two activities converge (i.e. merge) to one activity in forward pass as shown in fig-1, then EST of activity C = (greater EFT from activity A and B) + 1
- EFT = EST + duration -1

Forward pass is the direction from left to right as you traverse on network diagram. The example-1 below depicts EST (Early Start Time) and EFT (Early Finish Time) calculations below.

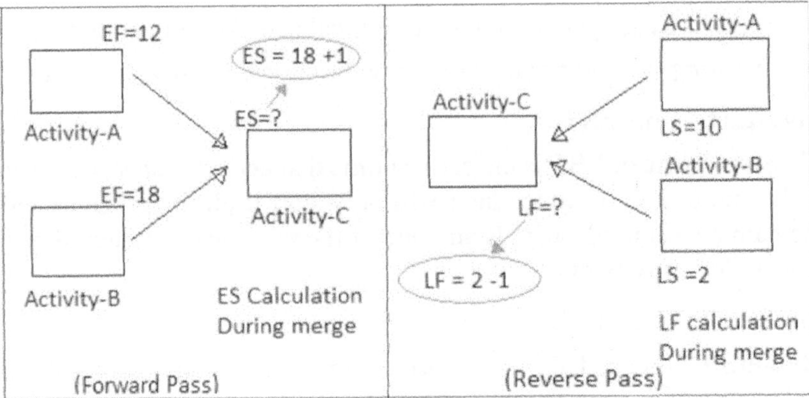

Tips for calculation of LST, LFT during reverse pass

- LFT of last activity = EFT of the same last activity
- For all other activities except the convergence mentioned in the next line, LFT = LST (of previous activity) -1
- If two activities converge(i.e. merge) to one activity in the reverse pass as shown in fig-2 then, LFT (of activity C) = (Lesser LST from activity A and B) -1
- LST = LFT - duration +1

Reverse pass is the direction from right to left as you traverse on network diagram. The example-2 below depicts EST (Early Start) and EFT (Early Finish) calculations below.

Example-1 EST (Early Start Time) and EFT (Early Finish Time)

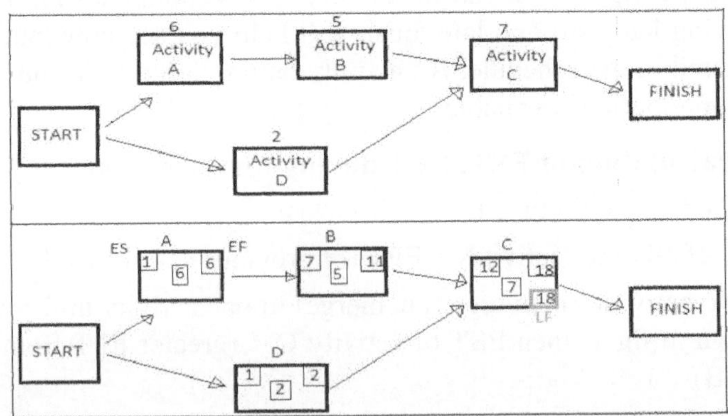

If one need to calculate EST and EFT from the above network diagram depicted in fig-3, follow tips mentioned above.

- EST of first activity is 1, EFT =EST+duration-1 = 6
- EST of activity-B = EFT of activity-A +1 ,
  Hence EST of activity-B = 6+1 = 7 ,
  EFT of activity B = 7+5-1 = 11
- EST of first activity D = 1
- EFT of Activity D = 1+2-1 = 2
- At activity-C two activities merge, hence we need to take bigger EFT from activity B(11) and D(2), hence EST = Bigger EFT +1 = 11 +1 =12
- EFT of activity C = 12+7-1 = 18

Example-2 LS (Late Start) and LF (Late Finish)

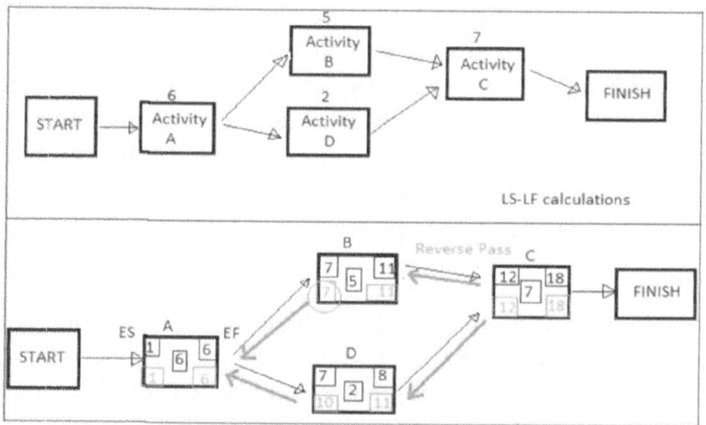

- LFT of activity C (last activity) = EFT of same activity =18
- LST of activity C = LFT - duration +1 = 18-7+1 = 12
- LFT of activity B and D = LST of C - 1 = 11 each
- LST of activity B = 11-5+1 =7
- LST of activity D = 11-2 +1 = 10
- Now LFT of activity A is lesser LST from B or D
  i.e. LFT of activity A = lesser LST -1 =7-1 =6
- LST of activity A = LFT-duration+1 = 6-6+1 = 1

Dummy activity: A dummy activity is intended to show a path of action in a project activity diagram and is employed when a logical relationship between two activities cannot be linked by showing the use of arrows linking one activity to another

Danglers: Dangling activities (also known as dangles) are loosely-tied activities in project schedules. They are activities with either open start dates or open end dates. All activities, except the first activity of a network, need to have a predecessor; otherwise, they will have open start dates.

Rules for Preparation of Networks

RULES FOR DRAWING NETWORK ANALYSIS

A fundamental ingredient in both PERT and CPM is the use of network systems as a means of graphically depicting a project. When a network is being constructed, certain conventions are followed to represent a project graphically, for it is essential that the relationship between activities and events are correctly depicted. Drawing a network diagram is a relatively easy task, and can be accomplished by listing each task on a piece of paper and representing the sequence in which the tasks take place. Lines and arrows are drawn between the pieces of paper to show which tasks follow on from others. The diagram aims to portray how the tasks relate to one another i.e. which tasks have to be completed before others begin and which tasks can be performed simultaneously. Creating the network is an iterative process and may involve a number of revisions before an optimum solution is found.

i. Sequencing

The initial step in project scheduling process is the determination of all specific activities that comprise the project and their interdependence relationships. In order to make a network following points should be taken into consideration.

- What job or jobs precede it?
- What job or jobs run concurrently?
- What job or jobs follow it?
- What controls the start and finish of a job?

Since all further calculations are based on the network, it is necessary that a network be drawn with full care. There are many ways to draw a network, in this lesson we will describe the method which follows the precedence table. The following example of preparation of Paneer (Cottage cheese) shows the basic steps required in drawing a network.

Example 1

For preparation of Paneer (Cottage Cheese) the following list represents major activities

1. Receive whole cow/buffalo milk
2. Standardize milk to obtain desired level of fat percentage
3. Take citric acid and prepare 1% solution
4. Heat the citric acid to 70 °C
5. Bring the standardized milk to boil on medium heat
6. Cool the milk to 70 °C and add slowly the solution of citric acid till yellowish whey separates
7. Strain the mixture through a clean muslin cloth.
8. Hold it under running water for a minute and then press out the excess water.
9. Hang the muslin for 15-20 minutes so that all the whey is drained out.
10. Prepare mould to form Paneer block
11. Fill the mass into the block and tie the muslin
12. Place it under something heavy for up to two hours
13. Cut the paneer into chunks and used as required.

Based on above list of different activities a precedence table may be formed which is given below;

Precedence table

Activity	Description	Preceding Activity
A	Receive whole cow/buffalo milk	-
B	Standardize milk to obtain desired level of fat percentage	A
C	Take citric acid and prepare 1% solution	-
D	Heat the citric acid to 70 °C	C
E	Bring the standardized milk to boil on medium heat	B
F	Cool the milk to 70 °C and add slowly the solution of citric acid till yellowish whey separates.	D,E
G	Strain the mixture through a clean muslin cloth.	F
H	Hold it under running water for a minute and press out the excess water.	G
I	Hang the muslin for 15-20 minutes & drain out all the whey.	H
J	Prepare mould to form Paneer block	H
K	Fill the mass into the block and tie the muslin	J
L	Place it under something heavy for up to two hours.	K
M	Cut the paneer into chunks and use as required.	L

In the above table due consideration has been given to precedings of an activity. While drawing the network, other factors will be considered. The activity A has no preceding activity and it is represented by an arrow line (above table).

Likewise activity C has no preceding activity and both activities A and C can be done simultaneously so they are shown as concurrent activities. Activities B and D are preceded by the activities A and C respectively. The complete network is shown in the given Figure

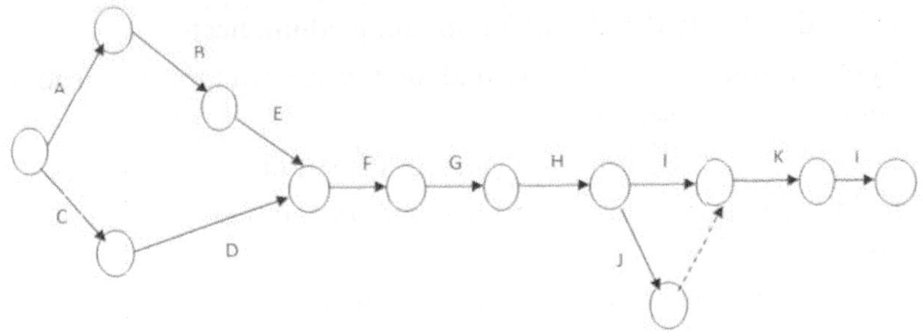

Network diagram

Guidelines for Drawing Network Diagram

There are number of rules in connection with the handling of events and activities of a project network which are given below:

a) Each activity is represented by one and only one arrow in the network. This implies that no single activity can be represented twice in the network. This is to be distinguished from the case where one activity is broken into segments. In such a case each segment may be represented by a separate arrow.

b) No two activities can be identified by the same beginning and end event. In such cases, a dummy activity is introduced to resolve the problem as shown in Fig. 19.2

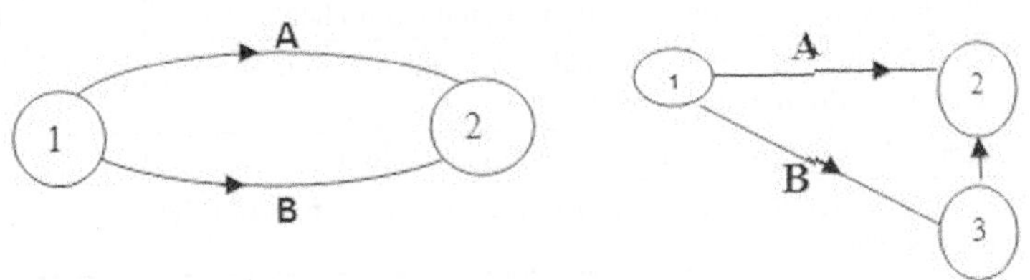

Using a dummy activity

c) In order to ensure the correct precedence relationship in arrow diagram following question must be checked whenever any activity is added to a network.

What activity must be completed immediately before this activity can start?

What activities must follow this activity?

What activities must occur simultaneously with this activity?

d) Thus a network should be developed on the basis of logical or technical dependence.

e) The arrows depicting various activities are indicative of logical precedence only; hence length and bearing of the arrows are of no significance.

f) The flow of the diagram should be from left to right.

g) Two events are numbered in such a way that the event of higher number can happen only after the event of lower number is completed.

h) Arrows should be kept straight and not curved. Avoid arrow which cross each other.

i) Avoid mixing two directions vertical and standing arrows may be used if necessary.

j) Use dummy activity freely in rough graph but final network should have only reluctant dummy.

k) The network has only one entry point called the start event and one point of emergence called end event.

l) Angle between the arrows should be as large as possible.

Error in Drawing Network

There are three types of errors which are common in network diagrams

i. **Dangling error:** To disconnect an activity before the completion of all activities in a network diagram is known as dangling.

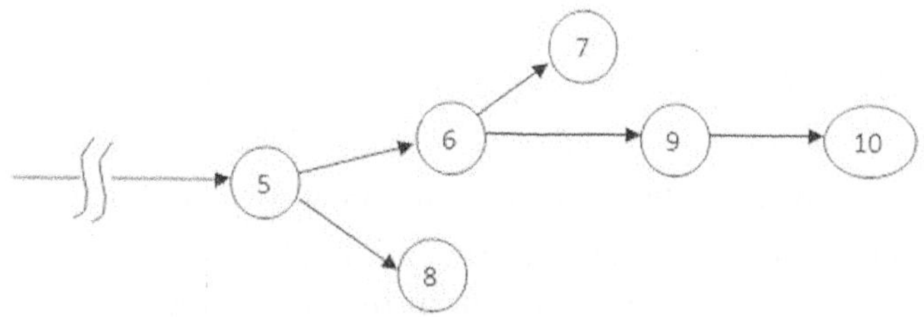

Dangling error

In Fig. 19.3 the activity 5 to 8, 6 to 7 are known as dangling error. These are not last activities in the network.

ii. **Looping error:** Looping error is also known as cyclic error in the network. Drawing an endless loop in a network diagram is known as error of looping as shown in Fig. 19.4

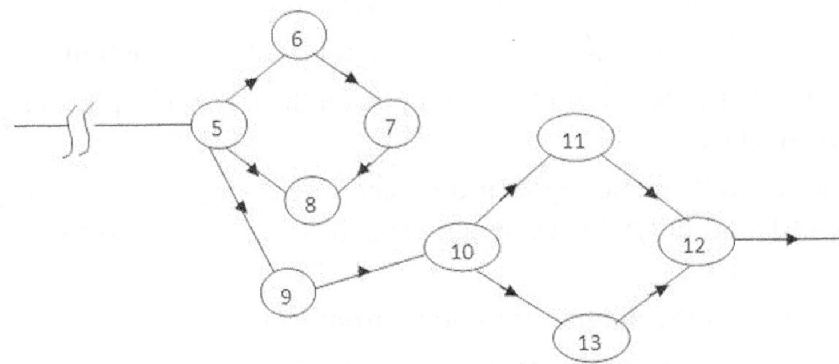

Appearance of a loop in the network

Reductancy error

Unnecessarily inserting the dummy activity in a network diagram is known as error of reductancy as shown in Fig 19.5 in which putting an dummy activity from 10 to 12 is a reductancy error.

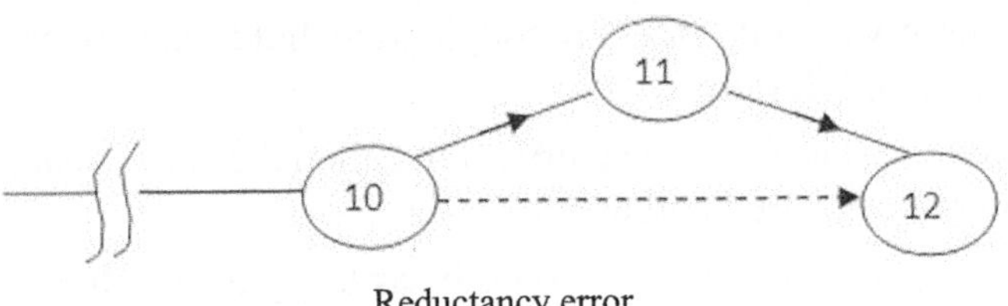

Reductancy error

Labeling of a Network Diagram

For network representation it is necessary that various nodes be properly labeled. For convenience, labeling is done on a network diagram. A standard procedure called i-j rule developed by D.R.F Fulkerson is most commonly used for this purpose.

Fulkerson□s i-j Rule

Step 1: First, a start event is one which has arrows emerging from it but not entering it. Find the start event and label it as number1.

Step 2: Delete all arrows emerging from all numbered events. This will create at least one new start event out of the preceding events.

Step 3: Number all new start events □2□, □3□ and so on. No definite rule is necessary but numbering from top to bottom may facilitate other users using the network when there are more than one new start event.

Step 4: Go on repeating step no. 2 & 3 until the end reached.

These rules are illustrated by taking into consideration the Example 19.1 and network diagram as shown in Fig.19.6

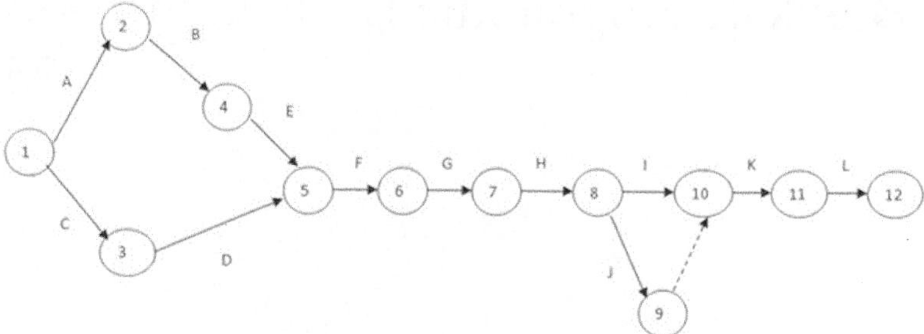

Network diagram of preparation of paneer

Block 4: Programme Evaluation Tools and Model

Unit 1

Programme Evaluation Tools and Model

An important decision that the management has to take during implementation of the process of evaluation is about which model to choose. The choice of a single model will mainly depend on the nature of the problem or the situation in an organization. However an integrative approach using more than one model may also be used. Before implementation of any of these models and before the actual process of evaluation starts it is important to make suitable preparations and create a conducive environment in the organization for successful conductance of the evaluation. The following points are to be taken in to consideration before and at the time an evaluation is carried out in an organization:

1. Bennett's Hierarchy of Evaluation

Introduction to Bennett's hierarchy – Background and description

Bennett's hierarchy has been used for almost 35 years in Cooperative Extension. Educators continue to relate well to this hierarchy in evaluating their Extension programs. Bennett's hierarchy contains seven sequential steps (input, activities, participation, reaction, knowledge, skills, opinions, aspirations-KASA, practice change, and end results/social, economic, environmental conditions-SEEC). The first four steps focus around process evaluation, while the last three steps focus on outcome/impact evaluation. Modifications were made to the hierarchy by ***Bennett and Rockwell in1995 and in 2000*** by adding a continuum linking program evaluation and program development. This revision helped educators understand that evaluation should be considered upfront in the design or planning phase of a program, not as an after-program activity. Bennett's Hierarchy is often shown as a staircase with inputs on the bottom and impact at the top. The idea being that you are climbing the staircase. I like to show it as radiating circles. The idea being that as you move up, your outcomes are stronger and further reaching.

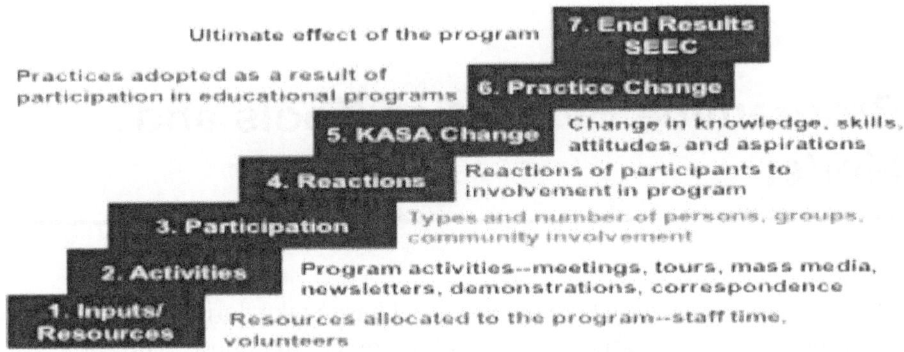

Fig 1: Bennett's Hierarchy of Evaluation

Kellogg Logic Model	Inputs	Activities	Outputs	Short-term Outcomes		Med-term Outcomes	Long-term Outcomes
Bennett's Hierarchy Logic Model	Inputs	Activities	Participation	Reactions	Knowledge Attitude Skills Aspirations	Practice Change	Impact

7. END RESULTS	Social economic, environmental and individual consequences of the program.
6. PRACTICE CHANGE	Adoption and application of knowledge, attitudes, skills, or aspirations.
5. KASA CHANGE	Knowledge – What do you know? Attitudes – How do you feel? Skills – What can you do? Aspirations – What would you desire?
4. REACTIONS	Degree of interest Like or dislike for activities Perceptions of projects
3. PEOPLE INVOLVEMENT	Number of people reached Characteristics of people Frequency & intensity of contact
2. ACTIVITIES	Newspaper or newsletter Articles Demonstrations Discussion groups Workshops
1. INPUTS	Staff time Costs Resources used

Fig 2: Working from 1-7 for planning & working from 7-1 for evaluation

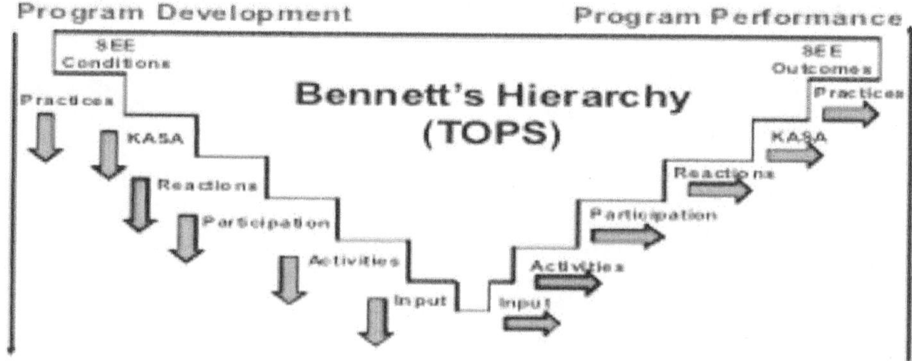

Fig 3: Bennett and Rockwell's TOPs Model

Today, in a time of continued reduction in government funding, Extension professionals are challenged more than ever before to document outcomes of programs and address stakeholder demands for accountability. This article provides a framework for linking Bennett's hierarchy to program outcomes and costs. Extension professionals could use this framework to link program outcomes and costs associated with such outcomes.

Examining Bennett's hierarchy from a different lens provides some insights to link the hierarchy with outcomes and costs associated with evaluating a program. A positive association between the seven steps of hierarchy and outcome type (short, intermediate, and long term) is proposed. As one moves up the hierarchy, the evidence of program impact gets stronger. Collecting evidence to assess impact of programs at the higher levels of the hierarchy becomes costly and time consuming and requires more skill. For example, one may use simple pre-post measures to assess short-term outcomes. On the other hand, to assess behavior/practice change, follow-up of participants is required, which adds to the cost of evaluating a program. This needs to be communicated to field-based educators so that they can plan early on what is needed in terms of costs, time, skills (data collection, analysis, and interpretation), and resources needed to evaluate an Extension program.

Linking Bennett's Hierarchy to Program Outcomes and Costs

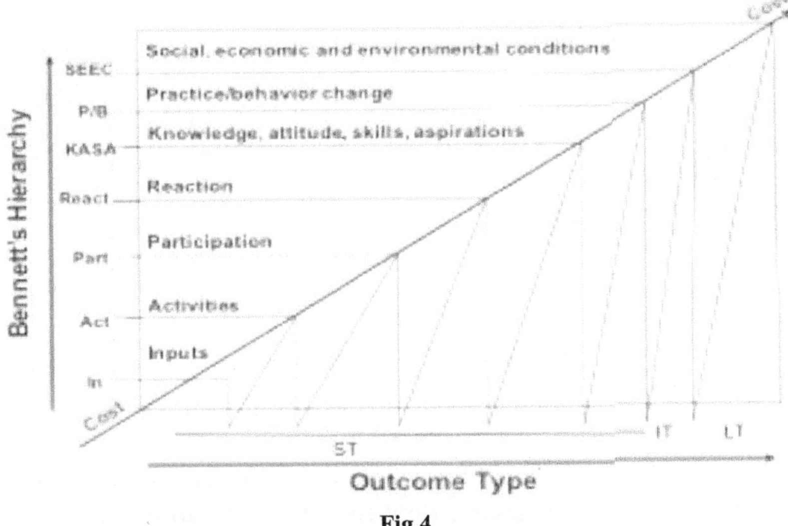

Fig 4

Based on the information presented in Figure 4, a framework for linking Bennett's hierarchy to program outcomes and costs was developed (Table 1). As shown in Table 1, evaluating programs at the lower levels (input, participation, activities, and reactions) may require little effort and are less expensive. However, evaluating at the lower levels will help program staff to assess ongoing program activities to make adjustments as the program progresses and to see whether or not the program is being implemented as planned. On the other hand, documenting and/or collecting evidence at higher levels of the hierarchy (KASA change, behavior change, and SEEC) requires skills relative to questionnaire development, data collection and analysis, interpretation, and reporting. In addition, it may also require understanding of evaluation designs, data collection at multiple points, sophisticated statistical analyses such as analysis of covariance, general equation modeling, use of covariates, etc. If these are planned in advance and properly done, the potential for program impact is stronger.

A Framework for Linking Costs and Program Outcomes Using Bennett's Hierarchy

	Process Evaluation				Outcome Evaluation		
Cost Outcomes	**Inputs**	**Activities**	**Participation**	**Reactions**	**KASA**	**Practice/ Behavior Change**	**SEEC**
Short Term	X	X	X	X	XX	XXX	-
Inter-mediate Term	X	X	X	-	XX	XXX	XXXX
Long Term	X	X	X	-	XX	XXX	XXXX
X = Low cost, effort, and evidence; XX—requires questionnaire development, data collection and analysis skills; XXX—requires understanding of evaluation designs, multiple data collection, additional analysis, skills, interpretation; XXXX—all of the above, time, increased costs, potentially resulting in stronger evidences of program impact.							

Steps for Using the Framework

i. First, decide on the level of evaluation you want to conduct, that is, process evaluation (lower levels of Bennett's) or outcome evaluation (higher levels of Bennett's) or both. In this step, develop a concept of linking program objectives to evaluation questions to outcomes.

ii. Second, identify key indicators for the evaluation. Make sure that the indicators you identified are measureable and relevant to the program.

iii. Third, once you decide on the type of evaluation and key indicators, consider all the costs that you might incur to conduct the evaluation (Figure 4 and Table 1). For example, one of your objectives is to assess practice/behavior change (higher levels of Bennett's) as a result of participation in a financial management program. This implies that you will be collecting data at least three times (pre, post, and delayed) to determine practice/behavior change. Perhaps you may also want to add a comparison group (non program participants). Including a comparison group will help make meaningful comparisons of a particular Extension program's strengths and weaknesses. In addition, you need to be cognizant of the time, labor, and other resources you need to have to carry out the evaluation at higher levels of the hierarchy.

iv. Fourth, design a matrix to document all the costs you incur to evaluate the program. Use of spreadsheets (Microsoft Excel) or other technology to document the costs for each level would be a good start.

Following these steps will help you to estimate the costs involved in evaluating an Extension program. In addition, these steps will also help guide the development of a budget for evaluation, especially if you are writing a grant for funding from a private or a public agency.

Advantages of Bennett's hierarchy of Evaluation

i. Clarity of Reporting Structure and Authority. ...
ii. Clear Pathway of Communication.

Disadvantages of Bennett's hierarchy of Evaluation

i. Hierarchies Can Stifle Collaboration. ...
ii. Lack of Innovation.

2. The Kirkpatrick Model of Training Evaluation

The Kirkpatrick model, also known as Kirkpatrick's Four Levels of Training Evaluation, is a key tool for evaluating the efficacy of training within an organization. This model is globally recognized as one of the most effective evaluations of training. The Kirkpatrick model consists of 4 levels: Reaction, learning, behavior, and results.It can be used to evaluate either formal or informal learning and can be used with any style of training. The Kirkpatrick Model has been widely used since **Donald Kirkpatrick** ***first published the model in the 1950s*** and has been revised and *updated 3 times* since its introduction. In 2016, it was updated into what is called the ***New World Kirkpatrick Model,*** which emphasized how important it is to make training relevant to people's everyday jobs. The various level of Kirkpatrick model of evaluation has been given below;

Level 1: Reaction

The first level is learner-focused. It measures if the learners have found the training to be relevant to their role, engaging, and useful.

There are three parts to this:

1. **Satisfaction**: Is the learner happy with what they have learned during their training?
2. **Engagement**: How much did the learner get involved in and contribute to the learning experience?
3. **Relevance**: How much of this information will learners be able to apply on the job?

Reaction is generally measured with a survey, completed after the training has been delivered. This survey is often called a 'smile sheet' and it asks the learners to rate their experience within the training and offer feedback.

Some of the areas that the survey might focus on are:

- Program objectives
- Course materials

- Content relevance
- Facilitator knowledge

Tips for Implementing Level 1: Reaction

- Use an online questionnaire.
- Set aside time at the end of training for learners to fill out the survey.
- Provide space for written answers, rather than multiple choice.
- Pay attention to verbal responses given during training.
- Create questions that focus on the learner's takeaways.
- Use information from previous surveys to inform the questions that you ask.
- Let learners know at the beginning of the session that they will be filling this out. This allows them to consider their answers throughout and give more detailed responses.
- Reiterate the need for honesty in answers – you don't need learners giving polite responses rather than their true opinions!

Level 2: Learning

This level focuses on whether or not the learner has acquired the knowledge, skills, attitude, confidence, and commitment that the training program is focused on.

These 5 aspects can be measured either formally or informally.

For accuracy in results, pre and post-learning assessments should be used.

Tips for Implementing Level 2: Learning

- Conduct assessments before and after for a more complete idea of how much was learned.
- Questionnaires and surveys can be in a variety of formats, from exams, to interviews, to assessments.
- In some cases, a control group can be helpful for comparing results.
- The scoring process should be defined and clear and must be determined in advance in order to reduce inconsistencies.
- Make sure that the assessment strategies are in line with the goals of the program.
- Don't forget to include thoughts, observations, and critiques from both instructors and learners – there is a lot of valuable content there.

Level 3: Behavior

This step is crucial for understanding the true impact of the training.

- It measures behavioral changes after learning and shows if the learners are taking what they learned in training and applying it as they do their job.
- It also looks at the concept of required drivers. That is, "processes and systems that reinforce, encourage and reward the performance of critical behaviors on the job."
- The results of this assessment will demonstrate not only if the learner has correctly understood the training, but it also will show if the training is applicable in that specific workplace.

This is because, often, when looking at behavior within the workplace, other issues are uncovered. If a person does not change their behavior after training, it does not necessarily mean that the training has failed.It might simply mean that existing processes and conditions within the organization need to change before individuals can successfully bring in a new behavior.

Tips for Implementing Level 3: Behavior

- The most effective time period for implementing this level is 3 – 6 months after the training is completed. Any evaluations done too soon will not provide reliable data.
- Use a mix of observations and interviews to assess behavioral change.
- Be aware that opinion-based observations should be minimized or avoided, so as not to bias the results.
- To begin, use subtle evaluations and observations to evaluate change. Once the change is noticeable, more obvious evaluation tools, such as interviews or surveys, can be used.
- Have a clear definition of what the desired change is – exactly what skills should be put into use by the learner? How is mastery of these skills demonstrated?
- Other questions to keep in mind are the degree of change and how consistently the learner is implementing the new skills. Will this be a lasting change?
- Evaluations are more successful when folded into present management and training methods.

Level 4: Results

This level focuses on whether or not the targeted outcomes resulted from the training program, alongside the support and accountability of organizational members.For each organization, and indeed, each training program, these results will be different, but can be tracked using Key Performance Indicators. Some examples of common KPIs are increased sales, decreased workers comp claims, or a higher return on investments.This level also includes looking at leading indicators. These are "short-term observations and measurements suggesting that critical behaviors are on track to create a positive impact on desired results."

Tips for Implementing Level 4: Results

- Before starting this process, you should know exactly what is going to be measured throughout, and share that information with all participants.
- If possible, use a control group.
- Don't rush the final evaluation – it's important that you give participants enough time to effectively fold in the new skills.
- It is key that observations are made properly, and that observers understand the training type and desired outcome.
- You can ask participants for feedback, but this should be paired with observations for maximum efficacy.
- Especially in the case of senior employees, yearly evaluations and consistent focus on key business targets are crucial to the accurate evaluation of training program results.

At all levels within the Kirkpatrick Model, you can clearly see results and measure areas of impact. This analysis gives organizations the ability to adjust the learning path when needed and to better understand the relationship between each level of training. The end result will be a stronger, more effective training program and better business results.

3. Jack Philip Roi Model Roi Means Return on Investment

As the name suggests the model focuses on whether the returns or outcome of a particular programme really match the investment made in them. This is extremely important as investment without positive outcomes is a loss to an organization and thus to be avoided. Thus this model can help us understand ROI in context of a particular programme. In the process of ROI analysis begins with deliberate attempts to isolate the effects of training on the data items. The ROI model requires a variety of data collection tools ranging from questionnaires and surveys to monitoring on the job performance. Different strategies have been used to accomplish the ROI calculation. The ROI formula is the annual net programme benefits divided by programme costs, and net benefits are the monetary value of the benefits minus the costs of the programme. It is as follows:

ROI (%) = Benefits – Costs × 100 costs

This model also recognizes that there would be intangible benefits that will be presented along with the ROI calculation organizations. Several common strategies began to emerge that can be considered best practices for calculating an ROI in training and development. The process of establishing evaluation targets has two important advantages.

i. It provides measurable objectives for the training staff to clearly measure progress for all programs or any segment of the process.
ii. Adopting targets focuses more attention on the accountability process, communicating a strong message to the training staff about the commitment to measurement and evaluation.

This model consists of five levels, they are discussed as follows:

a) **Reaction and Planned Action:** This level is similar to that of the Reaction level we discussed in Donald Kilpatrick Model. The reaction of the participants is recorded to understand the effectiveness of the programme. Besides the reaction the planned action or the way the participants intend to apply whatever learned by them is also taken in to consideration.
b) **Learning:** In this level stock is taken of the kind of learning or enhancement of knowledge occurring due to the programme.
c) **Job Applications:** This mainly focuses on whether the participants are applying what they learned while carrying out their job activity.
d) **Business Results:** This level is mainly concerned with whether the earlier step that is job application had produced any business results or positive outcomes for an organization.
e) **Return on Investment:** At this stage the formula can be applied to find out whether the outcome has exceeded the investments in terms of finances and other resources.

4. Cipp Model (Context, Input, Process and Product)

This model was developed by Stufflebeam et. al. in 1960s as a result of their experience of evaluating education projects for the Ohio Public Schools District. As the title suggests the model focuses on the four aspects of context, input, process and product. The model can be termed as decision-focused approach to evaluation. It emphasises the systematic provision of information for management and operation of a programme. This information in turn can be helpful in making certain decisions. This model mainly focuses on linking programme to decision making in order to ensure effective implementation of a programme. This is mainly done with the help of four aspects that is, context, input, process and product. These aspects are discussed in details as follows:

a) **Context:** In context the main focus is collecting data in order to conduct a needs assessment to determine goals, priorities and objectives of a particular programme. A context is created in order to understand the future course of action. The decision involved here is mainly with regard to planning.

b) **Input:** This deals with what is required for the programme to be successful, resources and strategies required for successful management of the programme. The decision here mainly focuses on the structure.

c) **Process:** This mainly highlights the actual process of implementation of the programme. Whether it is been implemented as per the plan and if there are any obstacle that need to be dealt with. The decision thus focuses on implementation.

d) **Product:** This aspect focuses on the outcome and whether the whole process of programme implementation was successful or not. Based on this decision can be taken with regard to the future course of action.

5. Kaufman's Five Levels of Evaluation

Kaufman, Keller & Watkins (1996) promoted an assessment strategy called the Organisational Elements Model (OEM) which involves four levels of analysis. Since the introduction of Kaufman's four-level OEM model, it has been widely used for evaluation. Kaufman, et. al. (1996), for example, later added levels of impact that go beyond the traditional four-level, training-focused approach which they felt did not adequately address substantive issues an organisation faces. Such modification to the model resulted in the addition of a fifth level, which assesses how the performance improvement program contributes to the good of society in general as well as satisfying the client. This model is mainly developed to evaluate a program from the trainee's point of view Let us discuss the **five levels** of analysis:

a) **Input and Process:** This analysis is based on inputs, and focuses on reductions of cost. This analysis can be divided in two parts, the first part focuses on the resources, financial, human resource and so on and the second part highlights the actual process and the reaction of the participants in the programme in order to understand the effectiveness of the programme.

b) **Acquisition:** This analysis focuses on the potentiality of the participants involved in the programme to grasp or acquire.

c) **Application:** This is in a way related to the previous analysis of acquisition. In this the actual application of knowledge and skills acquired is focused on.

d) **Outputs:** This relates to the products or services that are delivered to external clients.

e) **Outcomes:** This analysis highlights the value of the outputs (the aggregated products or services) delivered to external clients and their clients and ultimately to society.

6. Ciro Model (Content, Input, Reaction and Outcome)

The CIRO model was developed by Bird *et al.* The main elements of this model are Content, Input, Reaction and Outcome. It is very similar to the CIPP model. The main advantage of the CIRO model is that the objectives (context) and the training equipment (input) are taken in to consideration during the evaluation. The main elements of the model are discussed as follows:

a) **Context evaluation:** This evaluation is concerned with factors such as the effective identification of training needs and the setting of objectives with focus on organisations culture and climate. The context in which the programme is to be implemented is given utmost importance.

b) **Input evaluation:** This evaluation is concerned with the design and implementation of the training activity.

c) **Reaction evaluation:** It focuses on collecting and using information about the quality of trainees' experiences.

d) **Outcome evaluation:** It highlights the achievements gained from the activity. The activities are assessed at three levels: (a) immediate, (b) intermediate, and (c) ultimate evaluation. The focus is on the end results of the evaluation. Immediate evaluation attempts to measure changes in knowledge, skill, or attitude before a participant returns to the job. Intermediate evaluation refers to the impact of training on job performance and transfer of learning on the job. And ultimate evaluation attempts to assess the impact of training on departmental or organisational performance in terms of overall results.

Unit 2

Logic Framework Approach (LFA)

Introduction to LFA – Background and description

The **Logical Framework Approach** (**LFA**) is a methodology mainly used for designing, monitoring, and evaluating international development projects. Variations of this tool are known as **Goal Oriented Project Planning** (**GOPP**) or **Objectives Oriented Project Planning** (**OOPP**). The logical framework Approach is a method to develop the project's goals and activities and identify key information (such as risks) through a participatory approach. Basically, this means that instead of having a single person who designs the project all on his own, you seek the involvement of all concerned parties – also known as the stakeholders. The most important stakeholders are the potential beneficiaries of the project, or the target group.Getting the involvement of your target group and other stakeholders not only ensures that your project is well designed. It also makes sure that everybody is aware of the project and what it is supposed to change. It allows people to get involved, and develop a feeling of responsibility for the project and its results.This is known as developing ownership for the project's outcomes, and it is important to make sure that the positive realizations of the project are embedded locally and durable over time. People will care about the project and take care about its results.

Background

The Logical Framework Approach was developed in 1969 for the *U.S. Agency for International Development* (USAID). It is based on a worldwide study by Leon J. Rosenberg, a principal of Fry Consultants Inc. In 1970 and 1971, USAID implemented the method in 30 country assistance programs under the guidance of Practical Concepts Incorporated, founded by Rosenberg.It has been widely used by multilateral donor organizations, such as AECID, GIZ, SIDA, NORAD, DFID, SDC, UNDP, EC and the Inter-American Development Bank. Some non-governmental organizations offer LFA training to ground-level field staff. It has also gained popularity in the private sector, for example, in health care.

Project design with the Logical Framework Approach

The main steps in the Logical Framework Approach are:

- Getting to know the context
- Identifying the stakeholders
- Problem tree analysis (WS)
- Formulating the objectives tree (WS)
- Choosing the project's main strategy (WS)
- Formulating the logframe (WS)
- Verifying the project's design

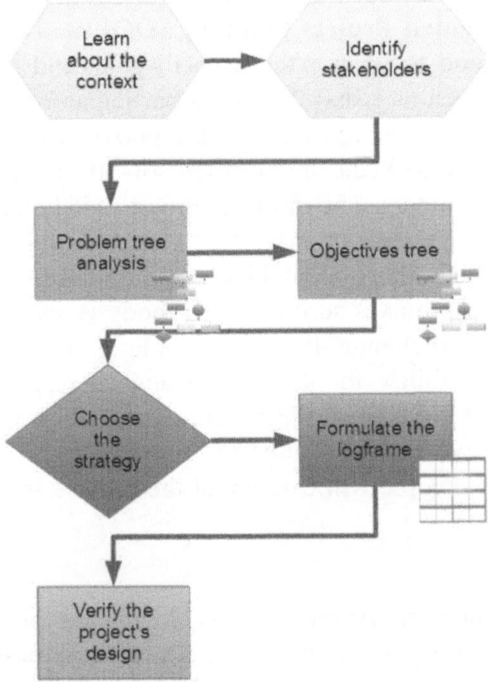

1. Learning about the context

Getting to know the context can be done in several ways:

- Reading reports, books, documents, websites
- Studying maps
- Talking with (local) experts
- Field visits
- Feasibility study

Things to learn about

- Is there a need for your specific expertise/assistance?
- Potential beneficiaries/target groups
- Potential intervention areas
- Potential for funding/financing
- Potential partners, networks
- Potential competition
- Potential impact of your interventions
- Political, social, economic situation of the country/region/area
- (Recent) history
- Current and potential market situation, economic stability
- Available human resources, level of education
- Gender situation, discrimination, social (in)stability
- Infrastructure, ICT, logistics
- Security situation
- Climatic conditions, natural risks (droughts, earthquakes, rains…)

With all this information as your background, it is time to identify who will be / has to be involved in the project.

2. Identify the stakeholders

Who are the people, groups, organizations, government bodies, companies… that are somehow involved or touched by this project? When you do a background analysis you will get a basic idea about who will be involved. These first contacts may lead you to identifying others. It is important to list the potential stakeholders, and then to determine who will get priority when you start analyzing the problems. Is the view of the local authorities more important than that of the ministry? Is the view of the local NGOs more important than that of the (local) government? Is the view of the farmers more important than that of the local trades people? Also, make sure you get the view of both men AND women. There are a number of tools that can help you with stakeholder analysis, such as mapping tools, Venn diagrams, organization charts (to identify administrative levels for instance) and so on.

When you've selected the most important groups, you can then analyse:

- Their main problems
- Their interests and needs
- Their strengths and weaknesses

- How they relate to other groups (cooperation/conflict; dependent/ independent)

This information will help you decide whose interests and view should get priority during the problem analysis.

3. Problem tree analysis

During the workshop (or sometimes it may be necessary to organize a series of workshops), you get a representative sample of the stakeholders that you identified with the group. **The problem tree analysis is an exercise that allows you to identify the different problems that people face, and the relationships between those problems.** The idea is to identify the core problem, and see what things are at the root cause of this central problem, and what other problems are a consequence of the core problem.

The first step is to identify the different problems

- Ask people to note different problems on cards: one card per problem.
- The problems have to be real – if there's doubt ask them to clarify by an example.
- Sometimes people will bring up things that they think are important for you – because you are rich and otherwise you're not going to give the money, right?
- Real problems also means that they are occurring now, not that they could occur if…
- A problem is not the absence of a solution – it's still too early to think of solutions. For instance, if someone thinks that he would get better crop yields if only he'd have fertilizers, the problem is not '*The absence of fertilizers*', but '*Poor (quality of the) soil*'.

4. Identify the core problem and establish a problem tree

Basically there are two ways to go about this. The group can agree on the core problem, and develop the problem tree around it. This means you look at what issues are causing the problem and which ones are a consequence of the core problem and you try to establish cause-and-effect relations between them. Often, the core problem is quite clear and just pops up. Also, your very presence and the fact that you can do specific things will influence the choice. If you are specialised in water and irrigation, the core problem will tend to be more 'watery' than when you are specialised in commercialization of food items.

The other way is to establish the cause and effect relationships between the problems first, and then select the core problem. This may give you more work to do, because it will be less clear from the onset what issues are less important or relevant than others.

The problem tree itself has roots and branches

The roots can be found below the core problem, these are the causes that lead to the issue that you›ve identified. Directly below the core problem you›ll find the cards with the most direct causes. Below these issues are their respective causes and so on.

The branches or the consequences are above the core problem. The most direct consequences can be found right above the core problem, then the issues that are a consequence of these direct consequences and so on.

When your tree is finished, you may find that there are still some gaps – meaning there are problems (cards) that you've not identified yet. Or maybe you don't understand the relation between a separate root/branch and the rest of the tree, and you may have to reflect on what is missing – or maybe there is a no relation at all.

The objectives tree

Now it's time to turn your problem tree into something constructive: **the objectives tree**. This simply means that you start at the top row of your problem tree, and rephrase every card in a positive statement or a solution.

- As before, sometimes you may need to rephrase the objective or clarify it.
- You also have to check the cause-and-effect relationships. If problem A causes problem B, does solution A also lead to the solution of B?
- Remove objectives that are not realistic.
- Sometimes it is necessary to add (intermediary) objectives.

5. Choosing the project's main strategy

There are often different ways to achieve the same objective. This means that it is important to agree on one single strategy. Often you'll find that there is one strategy that fits your organization's capabilities (type of activities, experience, available donor funds, capacities of staff and available human resources…) better than the other ones. Other elements that may influence your choice are:

- The price tag of each individual option;
- The potential risks and benefits;
- The ecological and financial impact and sustainability of each option;

- The potential impact on gender issues (both positive and negative);
- The social impact, including impact on social tensions and conflicts
- etc.

When there is more than one suitable strategy, you can use different criteria to select the best one.

6. Formulate the logical framework

Now it's time to put the information together in a logical framework.

- First you decide on the project's logic (intervention logic). This is the first column of the logical framework, and you can use the information from your objectives tree to identify the general objectives or goals of your project (the things above the main problem), the specific objective or purpose (the main problem itself) and the expected results (the things below the main problem).
- Then we jump to the far right column of the logframe, to identify the assumptions. These are the things that need to hold true in order for you to achieve your objectives.
- Finally it's time to think about how you are going to follow-up the progress of your project and evaluate its results. You'll need indicators for monitoring and evaluation (second column) and verification sources that specify how, where and when you can find that information (third column).

You can establish the basic version of your logframe during the workshop with your stakeholders. During the workshop, focus on getting the main ideas right and clear for everybody. Afterwards, you can put everything in nice sentences.

7. Identify the project logic

You now have the basic information needed to establish the **project's logic** – meaning the first column of the logical framework.

Generally the core problem, rephrased as an objective, becomes the purpose (or specific objective) of your project. This is the main reason why you started the project; the thing that you most urgently want to solve.

The solution of the core problem contributes (in the long run) to the solution of a problem in the larger society. This is the long term goal (or general objective) your project contributes to. Sometimes a project contributes to the solution of several issues that exist within that particular context and society.

The cards in the objective tree that were placed below the core problem/purpose – and that are part of the selected strategy – become the outputs and activities.

The outputs are the concrete things the team and beneficiaries will achieve during the life of the project. When the outputs are combined, the project's purpose should be achieved.

To achieve each output, you'll need a process that generally includes several activities. To do the activities, the project needs a number of means (inputs), such as staff, transport, tools, equipment, ICT, training, building materials, seeds… These come at a price, so now is a good time to get an indication of how much this all will cost.

The information from the objectives tree is necessary to develop the logical framework, but generally it takes some tinkering to perfect the logframe. Maybe some outputs and activities may have to be added. Or maybe the scope of the purpose has to be redefined, to make its achievement more realistic. During the workshop, make sure the project's basic logic is sound.

8. Identify risks and assumptions with your stakeholders

Because the problem tree has given you an insight in the relationship between different issues, it allows you to see what things may influence your main strategy and the achievement of your purpose. However, there may also be other elements that were not identified in the problem tree: generally you'll find the elements that come from the **context or environment**. You also have to take into account other risks such as **financial risks**, practical or **operational risks**, and **organizational risks**.

1. List the risks/assumptions
2. Eliminate those that are not important
3. Try to assess what the probability is that each risk occurs
    - If the risk is very likely to occur and the impact on the project is grave (it is doubtful you can achieve the project), then you have to redesign your project to eliminate or significantly reduce this risk. If this is not possible you should really think again about doing the project.
    - If the risk is likely to occur and the impact is important, but not life threatening, you should include it in the logframe and monitor the risk. If possible, you should try to influence the risk.
    - If the impact of the risk is low, you shouldn't include it into the logframe.
4. Identify whether the risk is a threat to one of the different outputs or to the achievement of the project's purpose itself. Other risks may threaten the sustainability of the results of the project, and therefore its long

term impact and contribution to the solution of some of the society's problems.

9. Indicators and their sources of verification

The next step in the workshop is to identify the indicators. There are a number of advantages to formulating the indicators with the whole group:

1. People that are closer to the field and the actual problems will know better how you can see that the situation improves. With the group, you can find **better indicators**.
2. It is easier to find **useful indicators** that can effectively be monitored.

When a project manager tries to identify indicators on his own, he or she will have a tendency to create indicators that can be expressed in numbers, such as the income per household to measure improved welfare. However, in the field it may be very difficult to get such a number, because people often don't have a notion of how money they earn. Besides, no-one likes to talk about how much they earn, do they? With a group you will be able to come up with indicators that are more easily measured. For instance whether people can afford to send their children to school, or if they can afford to improve their house or pay for transport.

When the indicators are identified, the group should check for each indicator:

- What information is needed
- Who should give/find that information
- In what form

If the indicator cannot be readily measured and verified, it should be replaced or dropped altogether.

10. Verify the project's design

After the workshop, it is important to finalise the project's design. This means that you have to check:

- **The project's** basic logic: are all the activities there to achieve the outputs that you want? Does the combination of the outputs lead to the purpose? Does the purpose contribute to the goal(s)? Aren't there any loose ends (missing activities, outputs that are not relevant for the purpose…)?
- Is the **target group clearly defined** (size, location, type of activities they do, gender, age…)?
- Who will receive the (financial/material) **benefits** from the project?

- Are all the elements **clear and unambiguous**? Will everyone concerned understand what is written in the same way? Aren't there any concepts that can be interpreted in different ways in various contexts (for instance: 'gender equality' is interpreted differently in Western Europe, in Eastern Europe, in Northern Africa, in Central Africa, in South-East Asia...)
- Are the project as a whole and its different elements **realistic**?
- Is the **analysis of the** risks/assumptions well made? Is it clear how you will react in the event that one or several of these risks occur?
- Is the monitoring system established (indicators + verification sources) ? Are the indicators well designed?
- Are the necessary **resources** there to execute the activities and to manage the project? What will you do with investments once the project is over?

11. Monitoring, reporting and evaluating with LFA

During the design phase, the LFA can be used to:

- Identify the stakeholders
- Identify issues and problems, determine how they are related in a problem tree and create a solutions tree
- Identify/formulate the goals, purpose, outputs and activities of the project
- Identify/formulate the risks and assumptions
- Identify/formulate the indicators and verification sources
- Create a detailed planning
- Establish the project's budget

Most of these elements can be found during a workshop (or a number of workshops) with stakeholders/partners/potential beneficiaries.

But apart from the identification and preparation of the project, the LFA can also be used during and after the execution of the project, to:

- Monitor
- Report
- Evaluate/review

The Logical Framework Matrix (Logframe)

The results of the stakeholder, problem, objectives and strategy analysis are used as the basis for preparing the Logical Framework Matrix. The Logical Framework Matrix (or more briefly the logframe) consists of a matrix with four columns and four (or more) rows, which summarise the key elements of a project plan and should generally be between 1 and 4 pages in length. However, this will depend on the scale and complexity of the project.Four-by-Four Grid – Rows from bottom to top (*Activities, Outputs, Purpose andGoal* & Columns representing types of information about the events (*Narrativedescription, Objectively Verifiable Indicators (OVIs)*) of these events taking place, Means of Verification (MoV) where information will be available on the OVIs, and Assumptions).

Project Description		Objectively verifiable indicators of achievement	Sources and means of verification	Assumptions
Goal	What is the overall broader impact to which the action will contribute?	What are the key indicators related to the overall goal?	What are the sources of information for these indicators?	What are the external factors necessary to sustain objectives in the long term?
Purpose	What is the immediate development outcome at the end of the project?	Which indicators clearly show that the objective of the action has been achieved?	What are the sources of information that exist or can be collected? What are the methods required to get this information?	Which factors and conditions are necessary to achieve that objective? (external conditions)
Outputs	What are the specifically deliverable results envisaged to achieve the specific objectives?	What are the indicators to measure whether and to what extent the action achieves the expected results?	What are the sources of information for these indicators?	What external conditions must be met to obtain the expected results on schedule?
Activities	What are the key activities to be carried out and in what sequence in order to produce the expected results?	**Means:** What are the means required to implement these activities, e. g. personnel, equipment, supplies, etc.	What are the sources of information about action progress? **Costs** What are the action costs?	What pre-conditions are required before the action starts?

Typical logical framework matrix. *Source*: BARRETO (2010)

How to Prepare The Logical Framework Matrix?

(Methodology adapted from BOND 2003; examples taken from EUROPEAN COMMISSION 2004)

A. First Stage — TOP DOWN

Project Description	Objectively verifiable indicators of achievement	Sources and means of verification	Assumptions
Goal			
Purpose			
Outputs			
Activities		Means and Costs	

First stage of the preparation of the logframe matrix. Source: BARRETO (2010)

- **Goal:** starting at the top and using the information from the Objective Tree write the overall objective of the project. The overall objective may be beyond the reach of this project on its own, for instances: "To contribute to improved family health and the general health of the rive ecosystem".
- **Purpose:** it describes the desired outcome that the project will achieve. This should be clear and brief. Example: "Improved river water quality".
- **Outputs:** describe the project intervention strategy. There may be several outputs. Example: "1) Reduced volume of wastewater directly discharged into the river system by households and factories".
- **Activities:** these are the tasks that are needed to achieve these outputs. There may be several for each output. Statements should be brief and with an emphasis on action words. Examples: "1.1) Conduct baseline survey of households and businesses; 1.2) Complete engineering specifications for expanded sewerage network, etc."
- **Inputs:** when required to do so provide additional information, such as the means and costs, which are needed to carry out these activities.

B. Second Stage — WORK ACCROSS

Project Description	Objectively verifiable indicators of achievement	Sources and means of verification	Assumptions
Goal			
Purpose			
Outputs			
Activities		Means and Costs	

Second stage of the preparation of the logframe Matrix. Source: BARRETO (2010)

- **Objectively verifiable indicators of achievement:** starting from the top to the bottom of the hierarchy of the objectives, begin to work across the logframe identifying the Objective Verifiable Indicators for measuring the progress in terms of quantity, quality and time. There are two kinds of indicators: 1. Impact indicators: related to the overall goal, helps to monitor the achievement and the impact of the project. Example: "Incidence of water borne diseases, skin infections and blood disorders caused by heavy metals, reduced by 50% by 2008, specifically among low income families living along the river". 2. Process (our outcome) indicators: related to the purpose and results. These measure the extent to which the stated objectives have been achieved. Example: "Concentration of heavy metal compounds (Pb, Cd, Hg) and untreated sewerage; reduced by 25% (compared to levels in 2003) and meets established national health/pollution control standards by end of 2007".
- **Sources and means of verification:** the source of verification should be considered and specified at the same time as the formulation of indicators. This will help to test whether or not the indicators can be realistically measured at the expense of a reasonable amount of time, money and effort. The SOV should specify how, who and when the information will be gathered.

C. Third Stage — BOTTOM UP:

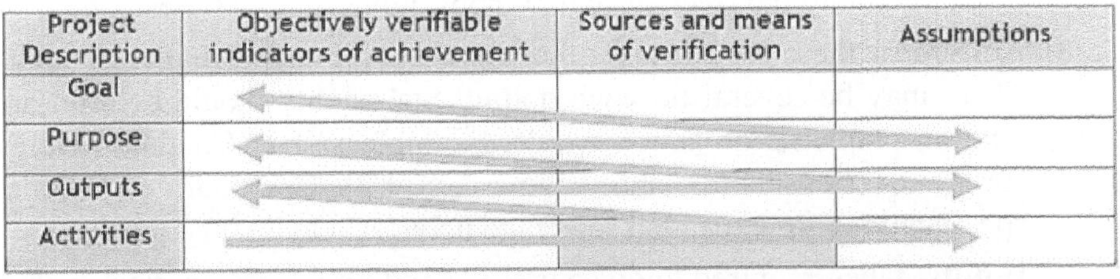

Project Description	Objectively verifiable indicators of achievement	Sources and means of verification	Assumptions
Goal			
Purpose			
Outputs			
Activities			

Third stage of the preparation of the logframe matrix. Source: BARRETO (2010)

- **Assumptions:** reflecting up from the bottom of the logframe, consider how, if each assumption holds, it will be possible to move to the next stage of the project. Assumptions are external factors that have the potential to influence (or even determine) the success of a project, but lie outside the direct control of project managers. Assumptions are usually progressively identified during the analysis phase. The analysis of stakeholders, problems, objectives and strategies will have highlighted a number of issues (i.e. policy, institutional, technical, social and/or economic issues) that will impact on the project 'environment', but over which the project may have no direct control. In the case of the river water pollution example, important assumptions might include issues related to: 1. Rainfall and river flow (beyond the project's control, but potentially critical in terms of changes in levels/concentration of pollutants found in the river); 2. Householders and businesses willingness to pay for improved sewerage connections.

Advantages of LFA

The advantages of using LFA are the following:

i. It ensures that fundamental questions are asked and weaknesses are analyzed, in order to provide decision makers with better and more relevant information.

ii. It guides systematic and logical analysis of the inter-related key elements which constitute a well-designed project.

iii. It improves planning by highlighting linkages between project elements and external factors.

iv. It provides a better basis for systematic monitoring and analysis of the effects of projects.

v. It facilitates common understanding and better communication between decision-makers, managers and other parties involved in the project.

vi. Management and administration benefit from standardized procedures for collecting and assessing information.

vii. The use of LFA and systematic monitoring ensures continuity of approach when original project staff are replaced.

viii. As more institutions adopt the LFA concept it may facilitate communication between governments and donor agencies. Widespread use of the LFA format makes it easier to undertake both sectoral studies and comparative studies in general.

ix. During initial stages, it can be used to test project ideas and concepts for relevance and usefulness

x. It guides systematic and logical analysis of the key interrelated elements that constitute a well-designed project (THE WORLD BANK 2000)

xi. It defines linkages between the project and external factors

xii. During implementation, the logframe serves as the main reference for drawing up detailed work plans, terms of reference, budgets, etc (WUR 2010)

xiii. A logframe provides indicators against which the project progress and achievements can be assessed (WUR 2010)

xiv. It provides a shared methodology and terminology among governments, donor agencies, contractors and clients (THE WORLD BANK 2000)

Disadvantages of LFA

i. Rigidity in project administration may arise when objectives and external factors specified at the outset are over-emphasised. This can be avoided by regular project reviews where the key elements can be reevaluated and adjusted.

ii. LFA is a general analytic tool. It is policy-neutral on such questions as income distribution, employment opportunities, access to resources, local participation, cost and feasibility of strategies and technology, or effects on the environment. LFA is therefore only one of several tools to be used during project preparation, implementation and evaluation, and it does not replace target-group analysis, cost-benefit analysis, time planning, impact analysis, etc.

iii. The full benefits of utilizing LFA can be achieved only through systematic training of all parties involved and methodological follow-up.

iv. Focusing too much on problems rather than opportunities and vision (WUR 2010)

v. Organisations may promote a blueprint, rigid or inflexible approach, making the logframe a straitjacket to creativity and innovation (THE WORLD BANK 2000)

vi. Limited attention to problems of uncertainty where a learning or adaptive approach to project design and management is required (WUR 2010)

vii. The strong focus on results can miss the opportunity to define and improve processes.

Block 5: Impact Assessment

Unit 1

Introduction to Impact Assessment

Introduction

Impact Assessment (IA) is a systematic process used to evaluate the potential or actual effects of a project, policy, or program on society, the economy, and the environment. It helps decision-makers understand the implications of their actions and ensures that interventions achieve their intended outcomes while minimizing negative consequences.

Meaning of Impact Assessment

Impact Assessment is a structured method of determining the significance of an action or intervention in various sectors. It involves analyzing both the intended and unintended effects of policies, projects, or programs. IA provides insights into the sustainability, feasibility, and effectiveness of an initiative before or after its implementation. By assessing social, economic, and environmental impacts, IA helps stakeholders make informed decisions.

Concept of Impact Assessment

The concept of IA is rooted in sustainable development and evidence-based policy making. It is used in multiple fields, including environmental management, social development, economic planning, and corporate governance. The key principles of IA include:

i. **Comprehensiveness**: Evaluating a wide range of impacts, including social, economic, and environmental factors.

ii. **Transparency**: Ensuring that the assessment process is open and inclusive.

iii. **Stakeholder Engagement**: Involving affected communities, experts, and policymakers in the assessment process.

iv. **Evidence-Based Decision Making**: Using data and research to guide policies and interventions.

v. **Sustainability Focus**: Ensuring that decisions promote long-term sustainability and do not cause harm to future generations.

Definition of Impact Assessment

Several definitions of Impact Assessment have been proposed by experts and organizations:

United Nations Development Programme (UNDP) defines IA as "a process of identifying the future consequences of a current or proposed action. It is used to ensure that policies, projects, and programs are socially, economically, and environmentally sustainable."

International Association for Impact Assessment (IAIA) describes IA as "the process of identifying, predicting, evaluating, and mitigating the biophysical, social, and other relevant effects of development proposals prior to major decisions being taken and commitments made."

World Bank defines IA as "an approach used to determine the effects of policies, programs, and projects on people, the economy, and the environment to improve outcomes and mitigate risks."

Aim of Impact Assessment

- Provide information for decision-making that analyses the biophysical, social, economic and institutional consequences of proposed actions;
- Promote transparency and participation of the public in decision-making;
- Identify procedures and methods for the follow-up (monitoring and mitigation of adverse consequences) in policy, planning and project cycles; and
- Contribute to environmentally sound and sustainable development.

Benefits and Advantages of Impact Assessment

1. There Is A Business Recovery Plan In Place

An impact assessment process produces a report that is based on data collected by various **impact** assessment methodologies. What this process achieves in an organization is beyond essential. The business will be in possession of information that details a business continuity plan in the event of a change in circumstances. This means that the business has a starting point instead of you trying to come up with strategies, or controls in the midst of the crisis.

2. Operational Advantages

Impact assessment can have the following effects on the operation and management of the organization:

- Improved productivity
- Improved customer service

- Cost reduction
- Improved product or service quality
- Better time management
- Improved planning and decision making
- Better management of resources
- Improved performance
- Better identification of legal and regulatory obligations

Additional organizational benefits also include;

- Business growth and development
- Strengthening of business partnerships, alliances and networks
- Improvement in business ideas and business plans
- Cost-leadership
- Increased market share
- Fostering united organizational culture by reducing resistance to change
- Better acceptance of developmental proposals
- Employee satisfaction and improved employee turnover rate
- Strengthening the organization's vision
- Encouraging more social, economic and environmental conscious decision making

Unit 2

Impact Assessment Indicators

Indicators for impact assessment

Meaning and concept of Indicator

The term indicator is often used interchangeably with data, targets, standards for evaluation, and even various modes of data collection, which undoubtedly inhibits a productive dialog on the subject. Such confusion is unwarranted, since the term has a precise connotation- in the literature on monitoring and evaluation of development. Simply stated, indicators are *variables designed to measure change in a given phenomenon or process*. They are analytical tools which facilitate the measurement of change and provide summary data for project design, implementation, and evaluation. Indicators are defined as "specific (explicit) and objectively verifiable measures of changes or results produced by an activity or intervention" (United Nations: 1985; p. 37). A few observations about indicators can be useful in further clarifying their nature. *First*, as the above definition implies, indicators provide quantitative data which can be analyzed using statistical techniques to produce estimates of a phenomenon. In cases when only qualitative information is gathered, it should be converted into numerical data for the purposes of analysis and presentation. *Second,* an indicator can capture one or more dimensions of a phenomenon or process depending upon the purpose of the study. We can, for example, construct an indicator for urbanization based on a wide range of variables such as number of school, higher educational institution, road connectivity, number of financial institution, health infrastructure, income etc. On the other hand, one can use each of the above items as a distinct indicator. Indicators focusing on different dimensions of a subject are difficult and more costly to construct. *Third*, several indicators can be used to capture the same phenomenon. For instance, direct indicators of food' consumption at the micro level include; 7-day recall of food purchases, 24-hour recall of food consumed, 7-day recall of food consumed, and weighing and measuring of the cocked food consumed. The availability of many indicators that can measure the same phenomenon poses both a challenge and opportunity to the evaluator. The task of choosing an appropriate indicator is dictated by several considerations. *Fourth*, it is useful to distinguish between direct or indirect (proxy) indicators.

The former involve the direct measurement of a phenomenon, for example, the measurement of per capita household income through income surveys. However, in many instances, the direct measurement of a phenomenon is not possible or cost effective. For example, because income surveys in rural areas often fail to provide reliable data, investigators tend use household asset, farm holdings, household expenditure, and level of living indicators as proxies for income. *Finally*, indicators must have a point of reference to determine the magnitudes of change, if any. Ideally, indicator data should be gathered at several points in time (before, during and after the project) to reveal the change or trends. One can then conclude, for example, that the income of a certain group of farmers increased by 40 percent at the end of the project or that it did not rise at-all. Depending upon the indicator, repetitive data collection may need to occur at specific intervals or during various seasons, for example rural unemployment during periods of slack agricultural labor demand, or rates of soil erosion during heavy rainfall periods. When time series data are not available, cross-sectional data can be used to make comparisons. In such situations, same or similar indicators are used to gather data for comparable groups or regions to determine change. Still in other situations, acceptable standards or targets can be used to measure progress or lack of it. In any case, the essential point is that indicator data is useful only baselines, standards, targets or other reference points against which changes can be examined, have been established. Indicators are usually classified in three categories depending upon the purpose for which they are used.

The technical or methodological requirements of selecting a good impact assessment indicator are listed below

i. **Valid:** An indicator should measure the phenomenon or process of interest

ii. **Reliable:** The conclusions based on an indicator should be the same if measured by different people at different times and under different conditions. In other words, they should be accurate.

iii. **Sensitive:** The indicator should be able to capture the change in the situation being observed. The issue of sensitivity is closely associated to precision.

iv. **Replicable:** We should be able to replicate an indicator in different projects and settings so that comparative analysis can be made. If the selected indicators cannot be used over t.ime in different settings, their value is undoubtedly limited.

v. **Availability of Data:** This means that the required data for an indicator should be easily accessible. There is little use of indicators if the data can not be gathered.

vi. **Costs of Data Collection:** Costs vary for different indicators depending upon the magnitude of information required, mode of data collection and scale of operation. One of the reasons indirect indicators are often used is to reduce cost. For example, anthropometric measures of nutrition, for example, are usually preferred because they are cost effective.

vii. **Availability of Technical and Organizational Resources:** Another practical consideration is the availability of technical and organizational resources. Indicators which require data which can be gathered with local resources in developing countries are preferable over others because the costs are reduced and the process of carrying out such efforts strengthens institutional capabilities in host countries.

viii. **Timeliness:** In the context of development projects, the rapid delivery of data on indicator is quite important. Indicators which meet these criteria better serve the information needs of project managers and the Agency.

ix. **Communicability:** Indicators should be simple to understand.

Types of impact indicators for technology and extension advisory services

i. **Social and behavioural indicators**

- Increases in the number of people reached
- Policy changes
- Changes in behaviour, e.g., adoption, attitude, perception and action
- Changes in community capacity;
- Changes in organizational capacity (skills, structures, resources);
- Increases in service usage
- Change in knowledge, skills and awareness level of the respondents/ clientele.

ii. **Socio-cultural indicators**

- Food security
- Poverty reduction
- Status of women improved
- Changes in resource allocation
- Changes in cash requirement

- Changes in labour distribution
- Nutritional implications

iii. Technology level assessment indicators

- Adoption of improved technology–symbolic and actual adoption
- Horizontal impact–increase in area under improved variety/breed
- Vertical impact–increase in productivity of improved variety/breed
- Reduction in cost of production
- Risk reduction
- Increase in annual income/economic capacity of the clientele
- Jobs or employment created

iv. Environmental impact assessment indicator

- Effect on soil erosion and degradation, silting, compact soil, soil contamination, water contamination
- Changes in hydrological regimes
- Effects on biodiversity, air pollution, greenhouse gases
- Forest area

v. Institutional impact assessment indicator

- Changes in organizational structure
- Change in the number of scientists
- Change in composition of the research team
- Multidisciplinary approaches and improvements
- Changes in funding allocated to the program
- Changes in public and private sector participation
- New techniques or methods

Unit 3

Approaches for Impact Assessment

Impact assessment approaches

Impact assessment approaches can be classified into four broad categories i.e. quantitative, qualitative, participatory and mixed methods.

i. Quantitative impact assessment approaches

- It focuses on assessing the degree and extent of the impacts quantitatively.
- Some degree of quantification may be necessary in all impact assessments, in order to evaluate the success of the Intervention and the extent of any adverse effects.
- Largely depends on micro-economic approaches following econometric models
- Involving baseline studies: The precise identification of baseline conditions, definition of objectives, target setting, rigorous performance evaluation and outcome measurement
- Costly and limited scope: Limited in the types of impacts which can be accurately measured and may pose difficulties for inference of cause and effect.

Types of Impact Assessments

A. **Based on time period of assessment:** Impact assessment may take place before approval of an intervention (ex -ante), after completion (ex post), or at any stage in between.

   i. **Ex-ante impact assessment:** Ex ante impact assessment is part of the need analysis and planning activity of the policy cycle. It involves doing a prospective analysis of what the impact of an intervention might be, so as to inform policymaking–the policymaker's equivalent of business planning. Forecasts potential impacts as part of the planning, design and approval of an intervention; Example–Assessment of adoption of Genetically Modified Rice variety.

   ii. **Ex-post impact assessment:** Ex post impact assessment is a part of the evaluation and management activity of the policy cycle.

Broadly, evaluation aims to understand to what extent and how a policy intervention corrects the problem it was intended to address. Impact assessment focuses on the effects of the intervention, whereas evaluation is likely to cover a wider range of issues such as the appropriateness of the intervention design, the cost and efficiency of the intervention, its unintended effects and how to use the experience from this intervention to improve the design of future interventions. Identifies actual impacts during and after implementation, to enable corrective action to be taken if necessary, and to provide information for improving the design of future interventions.

Example – Knowledge gain of trainees from a multimedia CD on SRI method of cultivating rice.

B. **Based on research design:** Impact assessment may be conducted using various designs

A. Experimental approaches

i. **Randomized evaluations:** In randomized evaluations, the programme benefits are extended to a randomly selected treatment group (beneficiaries), while keeping an identical group as control. The progress of the treatment and control groups on selected impact indicators is tracked over time (Khandker et al., 2010). Randomized evaluations are used when the eligible population is large enough to deliver the programme. A program needs to be gradually phased in until it covers the entire eligible population. The major advantage of this approach is its ability to avoid bias in selecting respondents.

ii. **Pre-test/Post-test with random assignment to intervention or comparison groups:** In these randomized experiments, study subjects are randomly assigned to a group that receives the technological intervention (study or treatment group) or a comparison group that does not receive the intervention (control or non-treatment group). Data for each group are collected before and after the intervention. At the end of the experiment, differences between the intervention and comparison groups can be attributed directly to the effect of the intervention, if the sample is large enough. Used in small samples consisting of less than 30 persons per group

iii. **Post-test only randomized experiment:** Two groups are randomly assigned the subjects and treatment conditions; Data collected only after the intervention; Statistical tests used are: Regression, t test, ANOVA.

B. Quasi-experimental approaches

i. **Pre-test/Post-test with non-random assignment to intervention or control groups**: In this design, data are collected before and after the intervention. Assigning subjects to the intervention and comparison groups is non-random. Comparison groups in the quasi-experimental design can be identified through propensity score matching. Propensity score matching – The control population (non-beneficiaries) is selected by 'matching' them with the actual beneficiaries on a few observable characteristics (personal and socio-economic attributes). The matched control groups can be selected either before project implementation (prospective studies) or afterwards (retrospective studies);

ii. **Two group post-tests only with non-random assignment**: In this data are collected only after the program has ended among participants who had received the Intervention and among non-participants. Matching participants and non-participants with similar characteristics and accounting for any relevant differences are especially important in the post-test only design to isolate effects of the intervention.

iii. **Double difference or difference-in-differences (DID) methods:** Estimates the effect of a specific intervention by comparing the changes in outcomes over time between a population that is enrolled in a program (the intervention group) and a population that is not (the control group). It can be applied in both experimental and quasi-experimental designs and requires baseline and follow-up data from the same treatment and control group. The mean difference between the 'after' and 'before' values of the outcome indicators for each of the treatment and comparison groups is calculated followed by the difference between these two mean differences. The second difference i.e., difference in the difference is the estimate of the impact of the program. This method is however used when the treatment group and the control group are heterogeneous at baseline. The y-axis represents the outcomes after a period of time (x-axis). Points B and D represented the observed outcomes of the Treatment and Control groups at the baseline of the Project intervention. Points E and C represented the observed outcomes of the Treatment and Control groups at the end of the project intervention. The first difference was measured by the difference in outcomes before-and-after for the treatment group which is given by O = D-E, while the second difference was measured by the difference in outcomes before-and-after for the control group given by W = B-C. The impact of the project intervention was therefore estimated using I = O-W.

Where,

A=Prior project start

B= observed outcomes of the Treatment at the baseline

D= observed outcomes of the control group at the baseline

E= observed outcomes of the Treatment at the end of the project/intervention

C= observed outcomes of the control group at the end of the project/intervention

U=unobserved changes without intervention

O=observed changes in the treatment

W=observed changes in the control

I=Impact of interest

- The first difference was measured by the difference in outcomes before-and-after for the treatment group which is given by;

  O = D-E

- The second difference was measured by the difference in outcomes before-and-after for the control group given by;

  W = B-C

- The impact of the Project was therefore estimated using

  I = O-W

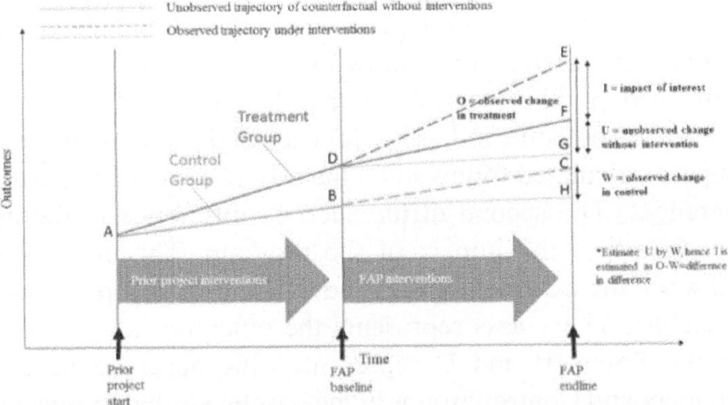

Conceptual diagram of difference-in-difference approach for impact estimation

iv. **Regression discontinuity design:** A pre-test–post-test comparison method design that elicits the causal effects of interventions by assigning a cut-off or threshold above or below which an intervention is assigned; In this design, the participants are assigned to intervention or comparison groups solely on the basis of a cut-off score on a pre-intervention measure. The average treatment effect is estimated by comparing observations lying closely on either side of the threshold. It is used in those conditions in which randomization is unfeasible and the researcher is interested in targeting an intervention or treatment to those who most need or deserve it.

Other methods are;

i. Propensity Score Matching
ii. Pipeline Methods
iii. Instrumental Variable (IV) Methods

C. Non-experimental designs (MLE, 2013)

These designs have only an intervention group without any control; Weakest designs for impact assessment. It is used under

- Limited resource condition
- Researchers are unable to create a comparison group
- When an intervention covers the entire population.

i. **Pre-test/post-test designs:** The researcher measures pre and post-intervention changes in the specific phenomenon (outcome indicator) between subjects; When changes occur in outcome indicators among the intervention participants, they cannot attribute all these changes to the intervention using this design alone because there is no comparison group.

ii. **Time-series designs:** The changes in outcome indicator over time are estimated to determine trends. The data is collected multiple times before and after the intervention to analyze trends before and after.

iii. **Longitudinal study:** The researcher records repeated measures of the same variables from the same people. A panel design is a special type of longitudinal design in which evaluators track a smaller group of people at multiple points in time and record their experiences in great detail.

iv. **Post-test only design:** Researchers observe the intervention group at one point in time after the intervention, focusing particularly on comparing responses of sub-groups based on such characteristics as age, sex, ethnicity, education or level of exposure to the intervention

D. Econometric Impact Assessment

i. **Partial Budgeting Technique:** This method is used when we have to calculate the economic impact of a small scale intervention or a single technology adoption. This method requires less data and allows early conclusions. In this method, if the profit remains same or decreases the intervention/technology is not more profitable than the technology used by farmer and therefore it should not be recommended. If the profit increases, the intervention/technology should be recommended to the farmers.

Profit/Loss= (Added returns -Added costs) + (Reduced costs - Reduced returns)

ii. **Net Present Value:** It is the present value of all benefit discounted at the appropriate discount rate, minus the present value of all costs discounted at the same rate.

NPV < 0, project/intervention is not profitable

NPV >0, project/intervention is profitable

$$NPV = \sum_{0}^{i} \frac{(Bt - Ct)}{(1+R)^1}$$

Where,

Bt is the benefit at time t

Ct is the cost at time t

r is the discount rate

t refers to the time period

iii. **Benefit Cost Ratio:** It is the ratio of the discounted benefits to the discounted costs of an investment with reference to the same point in time. Higher the BCR, better the intervention.

$$BCR = \frac{\sum Bt(1+r)^i}{\sum Ct(1+r)^t}$$

iv. **Internal Rate of Return or Return on Investment:** It is the rate at which NPV=0. IRR gives us a percentage value that indicates the profitability of the project. IRR is a way to compare between other similar options of investment (opportunity cost). If IRR is higher than the opportunity cost of investment in the project, it means the project has a positive impact.

$$IRR = r_1 + \frac{NPV_1}{NPV_1 - NPV_2}(r_2 - r_1)$$

r_1 = lower discount rate chosen

r_2 = higher discount rate chosen

NPV_1 = NPV at r_1

NPV_2 = NPV at r_2

v. **Economic Surplus Model:** To estimate economic impact, this model requires data related to technology targeted domain (production system, crops, inputs, outputs), yield and cost changes (increasing/stabilized), adoption level, R& D lag years, research and extension costs, market parameters (production, prices, elasticities) etc.

vi. **Adoption Quotient:** The extent of adoption was calculated by measuring the adoption quotient using the formula developed by Singh and Singh (1967);

$$AQ = \frac{\sum_{i-1}^{n} \frac{ei}{pi} X100}{N}$$

Where,

AQ = Adoption quotient

ei = Extent of adoption of each practice

pi = Potentiality of adoption of each practice

N = Total number of practices selected

ii. **Qualitative methods (Participatory methods):** it concerned with impacts within local socio-cultural and institutional context. Qualitative methods help to capture the things which are not possible to measure with the quantitative methods.

Reason for using qualitative methods to assess the impact in local socio-cultural and institutional context

i. Number reduce information
ii. Flexibility and reduce expenses
iii. People oriented research

The most widely used qualitative method of impact assessment is;

- In-depth interview
- Key informants rating
- Observation methods (Participants, Non-participants, Direct-Indirect, Obstrusive-Unobstrusive observation etc.)
- Case studies

- Document analysis
- Group discussion (Focus group discussion, Community meeting etc.)
- PRA/RRA (Venn diagram, Timeline analysis, Well-being ranking, Timeline analysis, Flow diagramme, Transect walk, Mobility map, Resource map, Seasonal calendar, Gender disaggregated task calendar etc.)
- Video and audio recording
- Client exit interviews
- Simulated Patient studies
- Artifacts

iii. **Mixed-methods:** it integrated the approach of both qualitative and quantitative methods in conceptual framework, design, analysis and interpretation so as to provide a comprehensive view of the impacts within the socio-cultural milieu. Mixed method used multiple methods to complement each other and at the same time triangulate the results to test its validity.

Unit 4

Environment Impact Assessment (EIA)

Introduction

Environmental Impact Assessment (EIA) is a process of evaluating the likely environmental impacts of a proposed project or development, taking into account inter-related socio-economic, cultural and human-health impacts, both beneficial and adverse.

UNEP defines Environmental Impact Assessment (EIA) as a tool used to identify the environmental, social and economic impacts of a project prior to decision-making. It aims to predict environmental impacts at an early stage in project planning and design, find ways and means to reduce adverse impacts, shape projects to suit the local environment and present the predictions and options to decision-makers.

Environment Impact Assessment in India is statutorily backed by the Environment Protection Act, 1986 which contains various provisions on EIA methodology and process.

Evolution & History of EIA

EIA is termed as one of the best policy innovations in the 1900s. The main aim of EIA is to conserve the environment and bring out the best combination of economic and environmental costs and benefits. Read the below-mentioned points to understand the Environmental Impact Assessment evolution and history:

- The birth of EIA is dated back to the 1970s.
- In 1969, The USA had brought its first National Environment Policy Act (NEPA) 1969.
- The EIA was initially practiced by developed nations but slowly it was also introduced in developing nations including India.
- Columbia and the Philippines are the earliest examples of developing nations who introduced EIA in their policies. Columbia brought it in 1974 while the Philippines in 1978.
- Worldwide, EIA is now practiced in more than 100 countries.

- By the mid-1990s, some 110 countries applied EIA as a major environmental policy.
- In 1989, EIA was adopted as the major development project by the *World Bank*.

History of EIA in India

- The Indian experience with Environmental Impact Assessment began over 20 years back. It started in 1976-77 when the Planning Commission asked the Department of Science and Technology to examine the river-valley projects from an environmental angle.
- Till 1994, environmental clearance from the Central Government was an administrative decision and lacked legislative support.
- On 27 January 1994, the then Union Ministry of Environment and Forests, under the Environmental (Protection) Act 1986, promulgated an EIA notification making *Environmental Clearance (EC) mandatory for expansion or modernization of any activity or for setting up new projects listed in Schedule 1 of the notification.*
- The Ministry of Environment, Forests and Climate Change (MoEFCC) notified **new EIA legislation in September 2006.**
- The notification makes it **mandatory for various projects** such as mining, thermal power plants, river valley, infrastructure (road, highway, ports, harbours and airports) and industries including very small electroplating or foundry units **to get environment clearance.**
- However, unlike the EIA Notification of 1994, the new legislation has **put the onus of clearing projects on the state government** depending on the size/capacity of the project.
- The other main laws in this regard are the Indian Wildlife (Protection) Act (1972), the Water Act (1974), the Air (Prevention and Control of Pollution) Act (1981), and the Biological Diversity Act (2002).

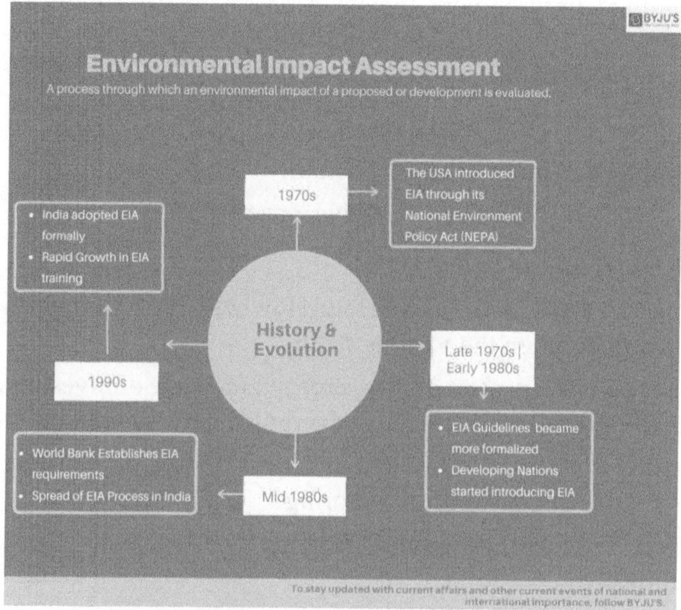

(*Source*: https://byjus.com/free-ias-prep/eia/)

Objectives of Environmental Impact Assessment

1. Identifying, predicting, and evaluating economic, environmental, and social impacts of development activities.
2. Providing information on the environmental consequences for decision making.
3. Promoting environmentally sound & suitable development by identifying appropriate alternatives and mitigation measures.

Importance of Environmental Impact Assessment

1. EIA is a good tool for prudent environment management.
2. It is government-policy that any industrial project in India has to secure EIA clearance from the Environment Ministry before approval for the project itself.

Current EIA Reports – India

EIA Notification 2020 draft has been made public. Once the EIA Notification 2020 will be published in the Official Gazette, it will replace EIA notification 2006. EIA has been in the news following EIA notification 2020 was drafted as one of the amendments will be the removal of public consultation from several activities (Put under Category B2). The important organization/agencies concerning EIA notification 2020 which aspirants should further read about:

1. Accredited Environment Impact Assessment Consultant Organization (ACO)
2. Central Pollution Control Board
3. Certificate of Green Building
4. Corporate Environment Responsibility
5. Eco-Sensitive Area/ Eco-Sensitive Zone

Steps in EIA process

EIA involves the steps mentioned below. However, the EIA process is cyclical with interaction between the various steps.

i. **Screening:** The project plan is screened for scale of investment, location and type of development and if the project needs statutory clearance.

ii. **Scoping:** The project's potential impacts, zone of impacts, mitigation possibilities and need for monitoring.

iii. **Collection of baselinc data:** Baseline data is the environmental status of study area.

iv. **Impact prediction:** Positive and negative, reversible and irreversible and temporary and permanent impacts need to be predicted which presupposes a good understanding of the project by the assessment agency.

v. **Mitigation measures and EIA report:** The EIA report should include the actions and steps for preventing, minimizing or by passing the impacts or else the level of compensation for probable environmental damage or loss.

vi. **Public hearing:** On completion of the EIA report, public & environmental groups living close to project site may be informed and consulted.

vii. **Decision making:** Impact Assessment Authority along with the experts consult the project-in-charge along with consultant to take the final decision, keeping in mind EIA and EMP (Environment Management Plan).

viii. **Monitoring and implementation of environmental management plan:** The various phases of implementation of the project are monitored.

ix. **Assessment of Alternatives, Delineation of Mitigation Measures and Environmental Impact Assessment Report:** For every project, possible alternatives should be identified, and environmental attributes compared. Alternatives should cover both project location and process technologies.Once alternatives have been reviewed, a mitigation plan should be drawn up for the selected option and is supplemented with an Environmental Management Plan (EMP) to guide the proponent towards environmental improvements.

x. **Risk assessment:** Inventory analysis and hazard probability and index also form part of EIA procedures.

Environmental Components of EIA

Depending on nature, location and scale of the project EIA report should contain all or some of the following components;

A. Air Environment

- Determination of impact zone (through a screening model) and developing a monitoring network
- Monitoring the existing status of ambient air quality within the impacted region (7-10 km from the periphery) of the proposed project site
- Monitoring the site-specific meteorological data, viz. wind speed and direction, humidity, ambient temperature and environmental lapse rate
- Estimation of quantities of air emissions including fugitive emissions from the proposed project
- Identification, quantification and evaluation of other potential emissions (including those of vehicular traffic) within the impact zone and estimation of cumulative of all the emissions/impacts
- Prediction of changes in the ambient air quality due to point, line and areas source emissions through appropriate air quality models
- Evaluation of the adequacy of the proposed pollution control devices to meet gaseous emission and ambient air quality standards
- Delineation of mitigation measures at source, path ways and receptor

B. Noise Environment

i. Monitoring the present status of noise levels within the impact zone, and prediction of future noise levels resulting from the proposed project and related activities including increase in vehicular movement.

ii. Identification of impacts due to any anticipated rise in noise levels on the surrounding environment

iii. Recommendations on mitigation measures for noise pollution

C. Water Environment

i. Study of existing ground and surface water resources with respect to quantity and quality within the impact zone of the proposed project

ii. Prediction of impacts on water resources due to the proposed water use/ pumping on account of the project.
iii. Quantification and characterization of waste water including toxic organic, from the proposed activity.
iv. Evaluation of the proposed pollution prevention and wastewater treatment system and suggestions on modification, if required.
v. Prediction of impacts of effluent discharge on the quality of the receiving water body using appropriate mathematical/simulation models.
vi. Assessment of the feasibility of water recycling and reuse and delineation of detailed plan in this regard

D. Biological Environment

i. Survey of flora and fauna clearly delineating season and duration.
ii. Assessment of flora and fauna present within the impact zone of the project
iii. Assessment of potential damage to terrestrial and aquatic flora and fauna due to discharge of effluents and gaseous emissions from the project
iv. Assessment of damage to terrestrial flora and fauna due to air pollution, and land use and landscape changes
v. Assessment of damage to aquatic and marine flora and fauna (including commercial fishing) due to physical disturbances and alterations
vi. Prediction of biological stresses within the impact zone of the proposed project
vii. Delineation of mitigation measures to prevent and / or reduce the damage.

E. Land Environment

i. Studies on soil characteristics, existing land use and topography, landscape and drainage patterns within the impact zone
ii. Estimation of impacts of project on land use, landscape, topography, drainage and hydrology
iii. Identification of potential utility of treated effluent in land application and subsequent impacts
iv. Estimation and Characterisation of solid wastes and delineation of management options for minimisation of waste and environmentally compatible disposal

F. Socioeconomic and Health Environment

i. Collection of demographic and related socio-economic data
ii. Collection of epidemiological data, including studies on prominent endemic diseases (e.g. fluorosis, malaria, fileria, malnutrition) and morbidity rates among the population within the impact zone
iii. Projection of anticipated changes in the socio-economic and health due to the project and related activities including traffic congestion and delineation of measures to minimise the adverse impacts
iv. Assessment of impact on significant historical, cultural and archaeological sites/places in the area
v. Assessment of economic benefits arising out of the project
vi. Assessment of rehabilitation requirements with special emphasis on scheduled areas, if any.

G. Risk Assessment

i. Hazard identification taking recourse to hazard indices, inventory analysis, dam break probability, Natural Hazard Probability etc.
ii. Maximum Credible Accident (MCA) analysis to identify potential hazardous scenarios
iii. Consequence analysis of failures and accidents resulting in fire, explosion, hazardous releases and dam breaks etc.
iv. Hazard & Operability (HAZOP) studies
v. Assessment of risk on the basis of the above evaluations
vi. Preparation of an onsite and off-site (project affected area) Disaster Management Plan

H. Environment Management Plan

i. Delineation of mitigation measures including prevention and control for each environmental component and rehabilitation and resettlement plan.
ii. Delineation of monitoring scheme for compliance of conditions
iii. Delineation of implementation plan including scheduling and resource allocation

Generalized EIA Process Flowchart

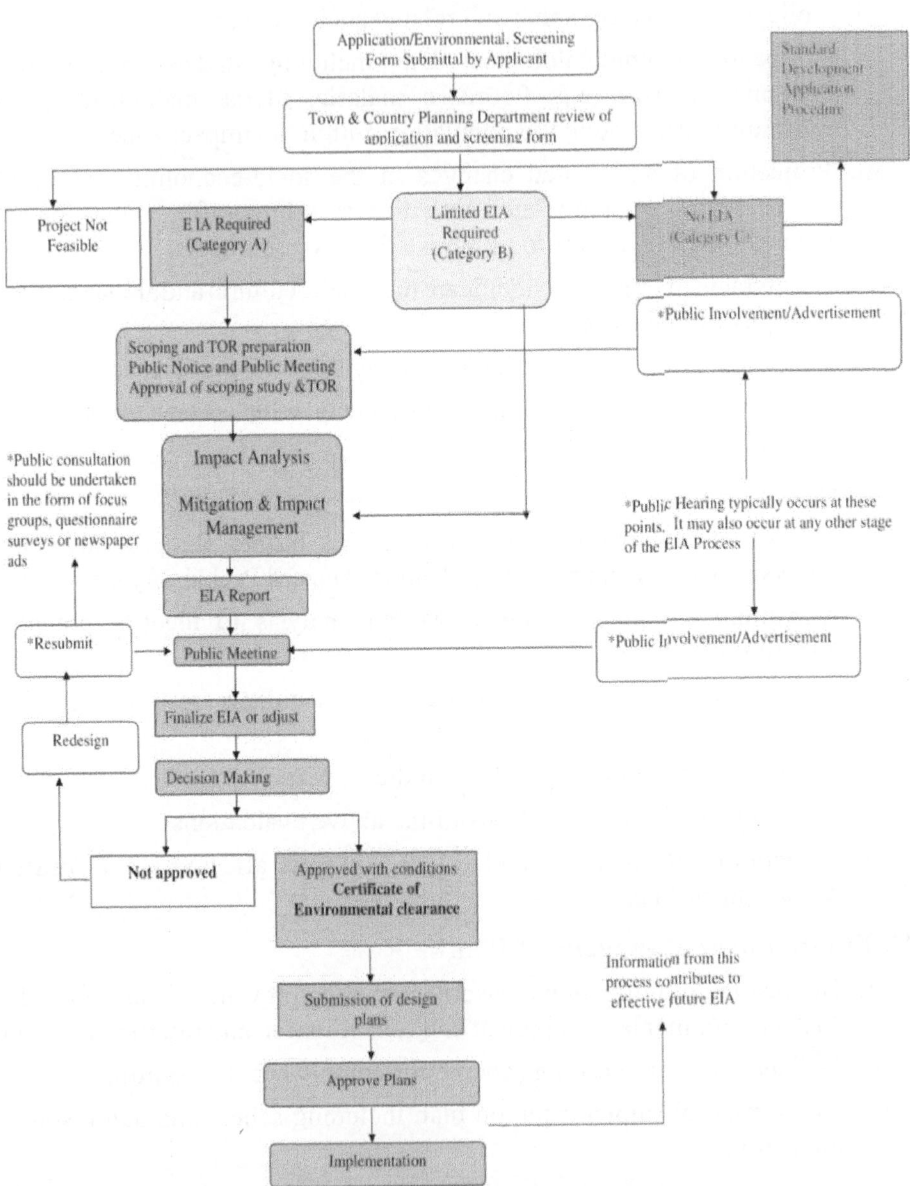

(*Source*: https://www.drishtiias.com)

Composition of the expert committees for EIA

The Committees will consist of experts in the following disciplines:

1. Eco-system management
2. Air/water pollution control
3. Water resource management
4. Flora/fauna conservation and management
5. Land use planning
6. Social Sciences/Rehabilitation
7. Project appraisal
8. Ecology
9. Environmental Health
10. Subject Area Specialists
11. Representatives of NGOs/persons concerned with environmental issues
12. The Chairman will be an outstanding and experienced ecologist or environmentalist or technical professional with wide managerial experience in the relevant development.
13. The representative of Impact Assessment Agency will act as a Member-Secretary.
14. Chairman and members will serve in their individual capacities except those specifically nominated as representatives.
15. The membership of a committee shall not exceed 15 members.

Stakeholders in the EIA Process

- Those who propose the project
- The environmental consultant who prepare EIA on behalf of project proponent
- Pollution Control Board (State or National)
- Public has the right to express their opinion
- The Impact Assessment Agency
- Regional centre of the MoEFCC

Salient Features of 2006 Amendments to EIA Notification

- Environment Impact Assessment Notification of 2006 has decentralized the environmental clearance projects by categorizing the developmental projects in two categories, i.e., **Category A (national level appraisal)** and **Category B (state level appraisal).**

- Category A projects are appraised at national level by Impact Assessment Agency (IAA) and the Expert Appraisal Committee (EAC) and Category B projects are apprised at state level.
- State Level Environment Impact Assessment Authority (SEIAA) and State Level Expert Appraisal Committee (SEAC) are constituted to provide clearance to Category B process.

After 2006 Amendment the EIA cycle comprises of four stages:

- Screening
- Scoping
- Public hearing
- Appraisal

Category A projects require mandatory environmental clearance and thus they do not undergo the screening process.

Category B projects undergoes screening process and they are classified into two types.

- **Category B1 projects (Mandatorily requires EIA).**
- **Category B2 projects (Do not require EIA).**

Thus, Category A projects and Category B, projects undergo the complete EIA process whereas Category B2 projects are excluded from complete EIA process.

Importance of EIA

i. EIA links environment with development for environmentally safe and sustainable development.

ii. EIA provides a cost effective method to eliminate or minimize the adverse impact of developmental projects.

iii. EIA enables the decision makers to analyse the effect of developmental activities on the environment well before the developmental project is implemented.

iv. EIA encourages the adaptation of mitigation strategies in the developmental plan.

v. EIA makes sure that the developmental plan is environmentally sound and within the limits of the capacity of assimilation and regeneration of the ecosystem.

Benefits of EIA

i. EIA links environment with development for environmentally safe and sustainable development.

ii. EIA provides a cost effective method to eliminate or minimize the adverse impact of developmental projects.

iii. EIA enables the decision makers to analyse the effect of developmental activities on the environment well before the developmental project is implemented.

iv. EIA encourages the adaptation of mitigation strategies in the developmental plan.

v. EIA makes sure that the developmental plan is environmentally sound and within limits of the capacity of assimilation and regeneration of the *ecosystem*.

Shortcomings of EIA Process

i. Applicability: There are several projects with significant environmental impacts that are exempted from the notification either because they are not listed in schedule I, or their investments are less than what is provided for in the notification.

ii. Composition of expert committees and standards: It has been found that the team formed for conducting EIA studies is lacking the expertise in various fields such as environmentalists, wildlife experts, Anthropologists and Social Scientists.

iii. Public hearing

- Public comments are not considered at an early stage, which often leads to conflict at a later stage of project clearance.
- A number of projects with significant environmental and social impacts have been excluded from the mandatory public hearing process.
- The data collectors do not pay respect to the indigenous knowledge of local people.

iv. Quality of EIA: One of the biggest concerns with the environmental clearance process is related to the quality of EIA report that are being carried out.

v. Lack of Credibility: There are so many **cases of fraudulent EIA studies** where erroneous data has been used, same facts used for two totally different places etc.

- Often, and more so for strategic industries such as nuclear energy projects, the EMPs are kept confidential for political and administrative reasons.
- Details regarding the effectiveness and implementation of mitigation measures are often not provided.
- Emergency preparedness plans are not discussed in sufficient details and the information not disseminated to the communities.

How to improve EIA process?

For effective EIA and report preparation, a project proponent and his consultants shall get hold of a variety of information. Some of the suggestions that might be useful are as follow:

Understand the Study Brief and environmental issues

- Get a thorough understanding of the study brief issued under the EIAO;
- Critically examine requirements for considerations of alternatives, if any; and any specific environmental concerns as listed down in the Study Brief; and
- Critically review any public comments received, particularly during the public exhibition of project profile.

Understand the Public Concerns

- Take proper account of public comments and opinions on environmental implications of the project;
- Get views from key stakeholders and green groups as early as possible;
- Understand the environmental issues - think from the perspective of the general public and those to be affected; and
- Present issues in layman terms.

Understand project history, the project justification & options available

- Identify and understand the issues, available options and solutions before deciding on whether detailed modelling or assessment is required;
- Critically examine the need and justification for a project. Collect background information from strategic, planning or feasibility studies; and

- Proactively look for environmentally-friendly alternative schemes or options to meet project need - avoidance versus mitigation or compensation. Laying down clearly the considerations being made to show the effort being given. Understand assumptions made in an EIA, scenarios development, input parameters in impact prediction

Know assumptions in EIA, especially basic information (e.g. traffic forecast & composition, pollution load, loss rate in dredging and etc.);

- Know assumptions made in identifying sensitive receivers, such as planned land uses, changing development programs, concurrent projects in the vicinity;
- Know spatial and temporal assumptions in construction methods and sequence, together with plant schedule. A project proponent shall critically review whether it is an engineering realistic construction program. Critical matters include, among others, whether a 24 hour operation is required for the works; adequacy of works areas; location of temporary haul road, concrete batching plant, rock crushing plant, jetty, mug-out portals, conveyor belts and etc.;
- Review practicality and programming of EIA recommendations and proposal of mitigation measures;
- Rigorously review the Implementation Schedule of Mitigation Measures in the EIA report and the EM&A Program;
- Review permanent structure/fixtures that are essential in operation phase such as location of ventilation shafts, passenger connections, overflow bypasses. Clarify roles of maintenance and implementation;
- Find out concurrent projects for cumulative impacts.

Case Study

- The MoEF constituted the **Western Ghats Experts *Ecology* Panel (WGEEP)** in 2010 under the Chairmanship of **Prof. Madhav Gadgil**.
- The Panel submitted its report in 2011 but it was not made public immediately due to its stringent assessment of the condition of Western Ghats.
- The report suggested many radical changes that needs to be brought to conserve Western Ghats.
- The recommendation if implemented would adversely affect mining mafia, sand mafia and local encroachers.
- Under pressure from various stakeholders, MoEF set up the **High**

Level Working Group (HLWG) under the Chairmanship of **Dr. K. Kasturirangan** to study recommendations of WGEEP.

- The HLWG had diluted many recommendations of WGEEP to satisfy the interests of various mafia.

WHAT NEED TO BE DOEN TO STRENTHEN THE EIA

i. **Independent EIA Authority.**
   - Sector wide EIAs needed.
   - Creation of a centralized baseline data bank.
   - Dissemination of all information related to projects from notification to clearance to local communities and the general public.

ii. **Applicability:** All those projects where there is likely to be a significant alteration of ecosystems need to go through the process of environmental clearance, without exception. No industrial developmental activity should be permitted in ecologically sensitive areas.

iii. **Public hearing:** Public hearings should be applicable to all hitherto exempt categories of projects which have environmental impacts. The focus of EIA needs to shift from utilization and exploitation of natural resources to conservation of natural resources. It is critical that the preparation of an EIA is completely independent of the project proponent.

iv. **Grant of clearance:** The notification needs to make it clear that the provision for site clearance does not imply any commitment on the part of the impact Assessment agency to grant full environmental clearance.

v. **Composition of expert committees:** The present executive committees should be replaced by expert people from various stakeholder groups, who are reputed in environmental and other relevant fields.

vi. **Monitoring, compliance and institutional arrangements:**
   - The EIA notification needs to build within it an **automatic withdrawal of clearance if the conditions of clearance are being violated** and introduce **more stringent punishment for noncompliance.** At present the EIA notification limits itself to the stage when environmental clearance is granted.
   - The composition of the NGT needs to be changed to include more judicial persons from the field of environment.
   - Citizen should be able to access the authority for redressal of all violation of the EIA notification as well as issues relating to non-compliance.

vii. **Capacity building:** NGOs, civil society groups and local communities need to build their capacities to use the EIA notification towards better decision making on projects.

References

Adrienne, M., Gundel, S., Apenteng, E. and Pound, B. (2011). Review of Literature on Evaluation Methods Relevant to Extension. Lindau, Switzerland: Global Forum for Rural Advisory Services, Lindau, Switzerland

American Evaluation Association (2018). https://www.eval.org/Portals/0/Docs/AEA%20 Evaluator%20Competencies.pd

Babioli, F. (2011). Indicators of Impact: A handbook of Impact Assessment, California, USA

Bagnol, B. (2014). Conducting participatory monitoring and evaluation. Decision tools for family poultry development. FAO Animal Production and Health Guidelines, Rome, Italy: FAO. Page.81-85.

Balakrishnan, R. Murai, A.S. Kalnar, Y. and Kumar, V. (2018). Techniques for Impact Assessment of Technologies. ICAR-Summer School on Advancements in Post Harvest Management of Legumes for Minimizing Losses and Sustainable Protein Availability. Page-252-259.

Balzer, L., Laupper, E., Eicher, V., and Beywl, W. (2020). The Key to Evaluation: 10 Steps – «evaluiert» in brief. Zollikofen: Swiss Federal University for Vocational Education and Training SFUVET. Downloaded from https://www.sfuvet.swiss/evaluiert

Bennett, C. (1975). Up the Hierarchy. Journal of Extension (March/April), Page.6-12.

Bennett, C.F. (1979). Analyzing impacts of extension programs. Department of Agriculture Washington, D.C., USA.

Bennett, C. (1975). Up the hierarchy. Journal of Extension [On-line], 13(2). Available at: https:// www.joe.org/joe/1975march/1975-2-a1.pdf

Bennett, C., & Rockwell, K. (1995). Targeting outcomes of programs (TOP): An integrated approach to planning and evaluation. Unpublished manuscript. Lincoln, NE: University of Nebraska.

Better Evaluation: Understand causes of outcomes and impacts, from (including a video from Jane Davidson) http://betterevaluation.org/plan/understandcauses

Boyle, R. and Le, Maire D. (1999). Building effective evaluation capacity: lessons from practice. New Brunswick, NJ: Transaction Publishers.

Bradford, R.W., Duncan, P.J. and Tarcy, B. (1999). Simplified Strategic Planning: A No-nonsense Guide for Busy People Who Want Results Fast. New York: Chandler House.

Braverman, M.T. and Engle, M. (2009). Theory and rigor in Extension program evaluation planning. Journal of Extension, 47(3). www.joe.org/joe/2009june/a1.php

Casley, Dennis, J. and Kumar, K. (1987). Project Monitoring and Evaluation in Agriculture". Baltimore: John Hopkin University Press.

Chen, H. (2012). Theory-driven evaluation: Conceptual framework, application and advancement. In: Strobl R., Lobermeier O., Heitmeyer W. (eds) Evaluation von Programmen und Projekten für eine demokratische Kultur. Springer VS, Wiesbaden

Chen, H.T. (2011). Practical program evaluation: Theory-Driven Evaluation and the Integrated Evaluation Perspective. Thousand Oaks, CA: Sage.

Chen, H.T. (2015). Practical program evaluation: Theory-Driven Evaluation and the Integrated Evaluation Perspective. Thousand Oaks, CA: Sage.

Dale, R. (2004). Evaluating Development Programmes and Projects, New Delhi, India: Sage Publications.

David, Mc., James, C. and Laura, R.LH. (2006). Programme Evaluation and Performance Measurement: An Introduction to Practice. New Delhi: Sage Publication Inc.

Duncan and Haughey (2017). SWOT Analysis. https://www.projectsmart.co.uk/swot-analysis.php.

Fetterman, D.M. (2012). Empowerment Evaluation: Learning to think like an evaluator. In M.C. Alkin (Ed.), Evaluation Roots (2nd edition) (pp. 304-322).

Fetterman, D.M. (2012). Empowerment Evaluation: Learning to think like an evaluator. In M.C. Alkin (Ed.), Evaluation Roots (2nd edition) (pp. 304-322). GFRAS. 2012. Guide to evaluating rural extension. Lindau, Switzerland: Global Forum for Rural Advisory Services (GFRAS).

Greene, J. (1988). Stakeholder participant and utilization in program evaluation. Evaluation Review, 12: 91–116.

Greene, J.C., Boyce, A., and Ahn, J. (2011). A values-engaged educative approach for evaluating education programs: A guidebook for practice. Champaign, IL: University of Illinois at Urbana-Champaign. http://comm.eval.org/communities/community-home/librarydocuments/ viewdocument? DocumentKey=f3c734c0-8166-4ba4-9808-a07e05294583

Greene, J.C., Boyce, A.S, and Ahn, J. (2011). Value-Engaged, Educative Evaluation Guidebook. University of Illinois, Urbana-Champaign.

Hall, A, Sulaiman, V.R, Clark, N. and Yoganand, B. (2003). From measuring impact to learning institutional lessons: An innovation systems perspective on improving the management of international agricultural research. Agricultural Systems, 78(2): 213–241.

Hind, J. (2010). Additionality: A useful way to construct the counterfactual qualitatively, Evaluation Journal of Australia, Vol.10, No. 1 http://www.fsnnetwork.org/sites/default/files/qualitative_approach_to_impact_evaluation.pdf

https://physicscatalyst.com/graduation/evaluation-in-education/

https://programs.online.american.edu/online-graduate-certificates/projectmonitoring/resources/evaluation-theory-and-practice

https://www.evalcommunity.com/career-center/evaluation-theories/

http://ecoursesonline.iasri.res.in/mod/resource/view.php?id=4384

https://www.researchgate.net/publication/230557767_Using_Programme_Theory_to_Evaluate_Complicated_and_Complex_Aspects_of_Interventions

https://archives.joe.org/joe/2010december/tt1.php

https://www.logframer.eu/book/export/html/16

https://byjus.com/free-ias-prep/eia/

https://www.drishtiias.com/to-the-points/paper3/environmental-impact-assessment-1/print_manually

Interaction: Annex 7: A Range of Quantitative, Qualitative and Theory-based Approaches for Defining the Counterfactualk http://www.interaction.org/annex-7-range-quantitative-qualitative-and-theory-basedapproaches-defining-counterfactual

Karthikeyan, C., Vijayaraghavan, K. and Lavanya, P. (2007). Formative evaluation of Kisan Call Centres. Tamil Nadu. Indian Journal of Extension Education, 43(1 & 2): 20-25.

Kumar, Krishna (1988). Assessing the Impact of Agricultural and Rural Development Assistance. A Background paper for a workshop on Impact Assessment of Agricultural and Rural Development Assistance. Washingtan, D.C.; Agency for International Development.

Management Systems International (1988). Current Practice and Immediate Needs for Collection and Presentation of Performance and Impact Data. A Report Submitted to PPC/CDIE/ORE.

Mayne, J. Contribution analysis: An approach to exploring cause and effect, ILAC Brief 16 http://www.cgiar-ilac.org/files/ILAC_Brief16_Contribution_Analysis_0.pdf

Miller, M.D., Linn, R. and Gronlund, N. (2009). Measurement and evaluation in teaching. New York: Merrill Press Edu Inc.

Mohr, L.B. (1999). 'The Qualitative Method of Impact Analysis', American Journal of Evaluation, 20 (1), 199k

Murray, P. (2000). Evaluating participatory extension programs: challenges and problems. Australian Journal of Experimental Agriculture, 40(4):519–526.

Narayan, D.(1993). Participatory Evaluation: Tools for Managing Change in Water and Sanitation (Technical Paper 207). Washington, D.C.: The World Bank.

Neuchatel Group. (2000). Guide for Monitoring, Evaluation and Joint Analyses of Pluralistic Extension Support. Lindau, Switzerland: Neuchâtel Group. www.g-fras.org/fileadmin/UserFiles/Documents/Frames-and-guidelines/M_E/Guide-forMonitoring-Evaluation-and-Joint-Analysis.pdf

Njuki, J, Mapila, M., Kaaria, S. and Magombo, T. (2008). Using community indicators for evaluating research and development programmes: Experiences from Malawi. Development in Practice, 18(4): 633–642.

OECD (1991), Principles for Evaluation of Development Assistance, Development Assistance Committee, OECD Publishing, Paris, https://www.oecd.org/development/evaluation/2755284.pdf

OECD (1998). Review of the DAC Principles for Evaluation of Development Assistance. Paris: DAC Working Party on Aid Evaluation. www.oecd.org/dataoecd/63/50/2065863.pdf (accessed 6 March 2024)

Patton, M.Q. (2013). Utilization-Focused Evaluation (U-FE) Checklist. Western Michigan University Checklists.

Rosanne, Lim. (2012). Why You Should Do a SWOT Analysis for Project Management.

Ross, P.H., Lipsey, M.W. and Howard, F. (2004). Evaluation: A Systematic Approach. New Delhi: Sage Publications Inc.

Rossi, P.H. and Freeman, H.E. (1985). Evaluation: a systematic approach (third edition). Beverly Hills, CA Sage Publications, Inc.

Radhakrishna, R. B., & Relado, R. Z. (2009). A framework to link evaluation questions to program outcomes. Journal of Extension [On-line], 47(3) Article 3TOT2. Available at: https://www.joe.org/joe/2009june/tt2.php

Radhakrishna, R. and Bowen, C.F. (2010). Viewing Bennett's Hierarchy from a Different Lens: Implications for Extension Program Evaluation. Journal of Extension, 48 (6):v48-6tt1.

Royse, David, Theyer, Bruce and Deborah, P. (2010). Programme Evaluation: An Introduction. USA: Wadsworth,Cengage Learning.

Sanders J. (1994). The program evaluation standards, 2nd edition. Joint committee on standards for educational evaluation. Thousand Oak, CA: Sage Publications, Inc.

Sasidhar, P.V.K. and Suvedi, M. (2015). Integrated contract broiler farming: An evaluation case study in India. Urbana, IL: USAID-MEAS. www.meas.illinois.edu (For Bennett's Hierarchy Example).

Sethuraman, P. Sivakumar, B.S., Sontakki, Sulaiman V.R. Saravanan, R. and Mittal, N. (2017). Manual on Good Practices in Extension Research & Evaluation. Agricultural Extension in South Asia (AESA), pp.1-14. http://www.aesa-gfras.net/

Shadish, W. R. Jr., Cook, T. D., & Leviton, L. C. (1991). Good theory for social program evaluation. Foundations of Program Evaluation: Theories of Practice (pp. 36-67). Newbury Park, CA: Sage. Newbury Park, CA: Sage.

Singh, K.M.P. and Singh, R.P. (1967). Ginger cultivation in Himachal Pradesh. Indian Farming, 30(11):25-26.

Srinath, L.S. (1975). PERT and CPM Principles and Applications, East-West Press, New Delhi.

Stern, E. (2015). Impact Evaluation, Prepared for the Big Lottery Fund, Bond, Comic Relief and the Department of International Development.

Suvedi, M. (2011). Evaluation of agricultural extension and advisory services-A MEAS training module. Urbana Champaign, IL: Modernizing Extension and Advisory Services Project. http://www.meas-extension.org/meas-offers/training/evaluatingextensionprograms

Suvedi, M. and Kaplowitz, M.D. (2016). Process skills and competency tools – what every extension worker should know–Core Competency Handbook. Urbana, IL: USAID-MEAS.

Suvedi, M. and Morford, S. (2003). Conducting Program and Project Evaluations: A Primer for Natural Resource Program Managers in British Columbia. Forrex-Forest Research Extension Partnership, Kamloops, B.C. Forrex Series 6.

Suvedi, M., Heinze, K. and Ruonavaara, D. (1999). How to Conduct Evaluation of Extension Programs. ANRECS Center for Evaluative Studies, Dept of ANR Education and Communication Systems, Michigan State University Extension, East Lansing, MI, USA https://msu.edu/~suvedi/Resources/Documents/4_1_Evaulation%20manual%202000.pdf

Sri Vidya College Of Engineering And Technology Course Material (Lecture Notes)

Tawfik, G.M., Dila, K.A.S., Mohamed, M.Y.F., Tam, D.N.H., Kien, N.D., Ahmed, A.M. and Huy N.T. (2019). A step by step guide for conducting a systematic review and meta-analysis with simulation data. Tropical Medicine Health, 47(46). https://doi.org/10.1186/s41182-019-0165-6.

United Nations ACC Task Force on Rural Development (1985). Monitoring and Evaluation: Guiding Principles". Rome; International Fund for Agricultural Development

United Nations Evaluation Group (2016). Evaluation Competency Framework. New York: UNEG. https://dictionary.apa.org/evaluator-credibility

USAID (2011). Evaluation policy. Washington, D.C., USA: Bureau for Policy and Planning.

Venkateswarlu, K. and Raman, K.V. (1993). Project Management Techniques for R&D in Agriculture. Sterling Publishers Pvt. Ltd., New Delhi.

Wholey, J.S., Harty, H.P. and Newcomer, K.E. (1994). Handbook of practical program evaluation. San Francisco, USA: Jossey-Bass Publishers.

Wollenberg, L. (2018). Impact evaluation methods: Qualitative Methods. Low Emissions Development Leader, CGIAR Research Program on Climate Change, Agriculture and Food Security (CCAFS) at the Green Climate Fund Independent Evaluation Unit Learning-Oriented Real-Time Impact Assessment (LORTA) Program Inception Workshop July 24-26, 2018 Bangkok, Thailand.